# FOOD POLICY TRENDS IN EUROPE
## Nutrition, Technology, Analysis and Safety

## ELLIS HORWOOD SERIES IN FOOD SCIENCE AND TECHNOLOGY

*Editor-in-Chief:* I. D. MORTON, Professor and formerly Head of Department of Food and Nutritional Science, King's College, London.
*Series Editors:* D. H. WATSON, Ministry of Agriculture, Fisheries and Food, and
M. J. LEWIS, Department of Food Science and Technology, University of Reading

**Food Biochemistry**   C. Alais & G. Linden
**Fats for the Future**   R.C. Cambie
**Food Handbook**   C.M.E. Catsberg & G.J.M. Kempen-van Dommelen
**Determination of Veterinary Residues in Food**   N.T. Crosby
**Food Policy Trends in Europe: Nutrition, Technology, Analysis and Safety**   H. Deelstra, M. Fondu,
    W. Ooghe & R. van Havere
**Principles and Applications of Gas Chromatography in Food Analysis**   M.H. Gordon
**Nitrates and Nitrites in Food and Water**   M.J. Hill
**Technology of Biscuits, Crackers and Cookies, 2nd Edition**   D.J.R. Manley
**Feta and Related Cheeses**   R.K. Robinson & A.Y. Tamime
**Vitamins and Minerals**   M. Tolonen
**Applied Human Nutrition: For Food Scientists and Home Economists**   A.F. Walker

*Forthcoming titles*

**Traditional Fermented Foods**   M.Z. Ali & R.K. Robinson
**Food Microbiology, Volumes 1 & 2**   C.M. Bourgeois, J.F. Mescle & J. Zucca
**Food Container Corrosion**   D.R. Davis & A.V. Johnston
**Technology of Meat and Meat Products**   J. P. Girard
**Dairy Technology**   A. Grandison, M.J. Lewis & R.A. Wilbey
**Separation Processes: Principles and Applications**   A. Grandison & M.J. Lewis
**Microbiology of Chilled and Frozen Foods**   W.F. Harrigan
**Modern Food Processing**   J. Lamb
**Food Technology Data**   M.J. Lewis
**Education and Training in Food Science: A Changing Scene**   I.D. Morton
**Modified Atmosphere Packaging of Food**   B. Ooraikul & M. E. Stiles
**Food: Production, Preservation and Safety, Volumes 1 & 2**   P. Patel
**Handbook of Edible Gums**   K.R. Stauffer
**Natural Toxicants in Food**   D.H. Watson
**Food Container Corrosion**   Wiese *et al.*
**Chilled Foods: A Comprehensive Guide**   Dennis & Shringer
**Food Process Modelling: Relating Food Process Design, Food Safety and Quality**   Niranjin & de Alavis
**Food Machinery**   Cheng

# FOOD POLICY TRENDS IN EUROPE
## Nutrition, Technology, Analysis and Safety

Editors

## HENDRIK DEELSTRA, MICHEL FONDU
## WILFRIED OOGHE and RUDY VAN HAVERE

Members of the Steering Committee of the International Symposium
*Food Policy Trends in Europe*, organized on the occasion of the
Centennial of the Belgian Food Regulations (1890–1990), held in
Brussels, October 22–24, 1990

**ELLIS HORWOOD**
NEW YORK  LONDON  TORONTO  SYDNEY  TOKYO  SINGAPORE

First published in 1991 by
**ELLIS HORWOOD LIMITED**
Market Cross House, Cooper Street,
Chichester, West Sussex, PO19 1EB, England

A division of
Simon & Schuster International Group
A Paramount Communications Company

Typeset in Times by Ellis Horwood Limited
Printed and bound in Great Britain
by Hartnolls, Bodmin, Cornwall


**Exclusive distribution by Van Nostrand Reinhold (International),**
**an imprint of Chapman & Hall, 2–6 Boundary Row, London SE1 8HN**

Chapman & Hall, 2–6 Boundary Row, London SE1 8HN, England

Van Nostrand Reinhold Inc., 115 5th Avenue, New York, NY10003, USA

Nelson Canada, 1120 Birchmont Road, Scarborough, Ontario M1K 5G4, Canada

Chapman & Hall Japan, Thomson Publishing Japan, Hirakawacho Nemoto Building, 7F,
1-7-11 Hirakawa-cho, Chiyoda-ku, Tokyo 102, Japan

Chapman & Hall Australia, Thomas Nelson Australia, 102 Dodds Street, South Melbourne,
Victoria 3205, Australia

Chapman & Hall India, R. Seshadri, 32 Second Main Road, CIT East,
Madras 600 035, India

*Rest of the world:*
Thomson International Publishing, 10 Davis Drive, Belmont, California 94002, USA


British Library Cataloguing in Publication Data

Food policy trends in Europe: Nutrition, technology, analysis and safety. —
(Ellis Horwood series in food science and technology)
I. Deelstra, Hendrik. II. Fondu, Michel. III. Ooghe, Wilfred. IV. Series.
368.8
ISBN 0–7476–0075–9


Library of Congress Cataloging-in-Publication Data

Food policy trends in Europe: nutrition, technology, analysis, and safety / editors, Hendrick
Deelstra, Michel Fondu, Wilfred Ooghe.
p. cm. — (Ellis Horwood series in food science and technology)
Includes bibliographical references and index.
ISBN 0–7476–0075–9
1. Nutrition policy — Europe. I. Deelstra, Hendrik. II. Series.
TX360.E8F66   1991
363.8′56′094–dc20

91–23025
CIP

# Table of contents

# Preface

In 1890, the first Belgian Food Act was voted by Parliament. This Act was one of the very first general food laws in Europe. In order to provide for the implementation of this Act, the Belgian Food Inspection Service was established in the same year. The year 1990 was the centenary of both these major events. To celebrate this anniversary various activities were scheduled, the most important of which were a lecture meeting and an international symposium.

The purpose of the lecture meeting, in which the Belgian Secretary of State for Public Health, the well-known French professor Debry, the Chairman of the Belgian Public Health Council and a senior EEC official representing EEC Commissioner K. Van Miert, took part, was to pay homage to the members of the Belgian Food Inspectorate.

The symposium, which was organized in conjunction with ILSI Europe, the Working Party on Food of the Flemish Chemical Society, the Federation of the Belgian Food Industry and the European Commission, was held at the Borschette Conference Centre in Brussels. It was presided over by Mr R. Van Havere, head of the Belgian Food Inspectorate. The aim of the symposium was to examine food policy developments in tomorrow's Europe. All food policy partners participated in the meeting: national and international authorities, scientists, economic and trading partners, and consumers. The symposium offered an ideal opportunity for all these partners to exchange ideas on the food policy of today and how it should develop, especially after the single European market has been achieved in 1993.

During the opening day, the various topics to be discussed at the symposium were introduced by representatives of an international organization, a scientific body and industry. This mental exercise focused on four main items: nutrition, food science and technology, food control and analyses, and safety assessment, each of which was dealt with in a separate session. A group of eminent international specialists gave their point of view on these items. Each session ended in an interesting discussion.

The last day of the symposium was introduced by the *rapporteurs* of the different sessions, who summarized the conclusions of the various sessions. In a round-table discussion, representatives of the various food chain partners explained their views on food policy trends in Europe.

Such a wide variety of topics was discussed during these fruitful days that it is almost impossible even to summarize them here. Legislation, analyses, manufacturing, hygiene, nutrition, novel foods, contaminants, additives, labelling, toxicity, safety, and many other items will have to be well balanced in the food sector of the single market in tomorrow's Europe.

During the symposium a poster session was held, featuring posters from scientists from all over Europe and giving details of subjects related to the topics discussed at the symposium.

The proceedings of the symposium contain most of the speeches and offer a wealth of data and viewpoints, which will help all food policy partners to understand each other and to achieve a transparent though safe food trade.

The Belgian Food Inspection Service

# Part I
# *Opening session*

# I.1

## The perspective to 1992

**P. S. Gray**
Head of Division Foodstuffs, Commission of the European Communities†

In art, the concept of perspective was developed extensively by the Flemish school of painters, many of whom lived and worked in the country we now know as Belgium. Good perspective involves the artist in painting within a coherent framework where, observing from a clearly defined position, he designs his picture in an ordered way, the framework converging at a definite point which may be outside of the picture that he is painting. Many of the old Flemish masters used a tiled floor which had a clear geometry to add structure and depth to their pictures. In food legislation the starting point for each Community Member State was of necessity different, and although their aims were similar they were not always identical and therefore, within the concept of a single market, the perspectives were confused. This paper, in analysing the perspective to 1992 within the framework already set out, seeks to look beyond that date to see what is necessary to achieve a coherent design for food law in Europe.

It is impossible to understand the role of food law in a modern open democratic society without reflecting on the nature of food itself. Food is, on the one hand, the most essential of all goods, the one which is vital for our very existence. Wherever we are, at home or at a restaurant, we have to eat what is on offer and our style of eating is often dictated by our social habits or economic circumstances. It is therefore of the utmost importance that food law should offer a high level of protection in terms of public health guaranteed by effective control.

On the other hand, food is a medium for creative expression through which is presented the diversity and richness of our European culture. Culinary art can manifest itself in simple traditional skills or in complex creative innovation. If we are to preserve this cultural heritage a European food law must allow for diversity and must not inhibit development, both in the industrial sense and in the adaptation of food to the needs of our changing society. The path of information rather than legal

† The views expressed in this paper are those of the author and do not necessarily represent those of his institution.

prescription in matters other than safety is the only way in which diversity and liberty of choice can be preserved.

Early food law, being an embodiment of guild traditions, tended to protect the 'state of the art' rather than the consumer. It was only with the growth of urban society and the development of an industrialized scale of food production that a more systematic approach to food law was developed. In the middle of the last century, food was frequently adulterated deliberately. The addition of water to beer and milk were commonplace, even sawdust and sand were used to 'extend' foods, but more sinister adulterants such as copper, lead, mercury and even arsenic compounds as colouring matters in a variety of products became so prevalent that in the United Kingdom in the late 1850s a committee was set up to investigate these frauds. In 1860 the Food & Drink Act was passed, making it an offence to sell food containing substances injurious to health. The big problem, then as now, was enforcement, and the lack of qualified official analysts meant that the law was not as effective as was hoped.

The Food and Drugs Act of 1875 laid down for the first time that '. . . no person shall sell to the prejudice of the purchaser any article of food or anything which is not of the nature, substance or quality demanded by such purchaser . . .' [1]. In the same year the Society of Public Analysts was formed, which made a major contribution to ensuring that sufficient qualified analysts were available to enforce the law. The experience gathered in the United Kingdom was certainly of great value in 1890 to the founding fathers of the first Belgian Food Law.

The development of modern food law has gone hand in hand with that of food science. The use of chemical preservatives in food has always been of concern to food scientists and in June 1923 a paper dealing with their use in foods was given at the fourth meeting of the International Union of Pure and Applied Chemistry held in Cambridge, UK. It stated that '. . . the trend of public and scientific opinion at the present time is towards the complete elimination of chemical preservatives from foods . . .' [2].

This spirit was in the minds of the pioneers of European food law when the first food directive which was adopted by the Council of Ministers in 1962. This directive, which deals with colours, has no EEC number because at that time no formal legal system for numbering had been worked out. There was great euphoria amongst Commission and Council officials not only for having fathered the first EEC directive but also for having reduced the number of food colours permitted in the (then six) Member States by about 40%. The adoption had been a triumph not only for diplomacy but also for consumer protection.

The initial approach of the Commission was based on the concept that a national law needed a Community law in order to ensure the free circulation of goods in the Community. For many years Community food legislation pursued a path dictated by this approach using article 100 of the EEC Treaty which called for unanimity. Progress was slow and, following a number of cases in the European Court of Justice, the first of which was in 1975 [3], followed by the famous Cassis de Dijon case in 1979 [4], the Commission put forward a new approach to food law. In the Communication of 8 November 1985 [5], it stated that the legislative approach followed in the past needs to be revised by drawing a distribution between, '. . . on the one hand, matters which, by their nature, must continue to be the subject of legislation and, on the other hand, whose characteristics are such that they do not need to be regulated ...' (point 7). Community legislation on foodstuffs should be limited to '... provisions justified by the need to:

— protect public health;
— provide consumers with information and protection in matters other than health and ensure fair trading;
— provide for the necessary public controls ...' (point 9).

The Communication went on to say that '... it is neither possible nor desirable to confine in a legislative straitjacket the culinary riches of twelve European countries ...'.

This approach has been mistakenly, and in some cases misleadingly, interpreted as applying the 'lowest common standards' which will in turn lead to a downward spiral of food quality. Nothing could be more mistaken. It is not a case of applying the 'minimum rules' but of applying the 'necessary rules', and applying them more strictly than in the past. The result, for the consumer, should be greater security and more choice and, for the ethical manufacturer, the elimination of malpractice which can, as the result of well-publicized food scandals, rebound also on the sales of products which are of irreproachable quality but unfortunately belong to the same category as those in which the scandals occurred. The guarantee against the downward spiral is to be found in paragraphs 3 and 4 of Article 100A of the EEC Treaty inserted by the Single European Act.

Shortly after this Communication, the Commission proposed to the Council a number of framework directives dealing with these essential requirements, the majority of which are now adopted. At the same time the Commission requested a wide delegation of powers for the enacting of implementing legislation and this has been granted to a large extent.

However there are three topics which have yet to be dealt with by general measures at Community level and which could constitute barriers to trade since Member States could invoke Article 36 of the Treaty in respect of them: namely novel foods, contaminants and food microbiology.

A draft directive has been in preparation for some time on novel foods. These are food ingredients which have never been marketed as such before and are not food additives since they are not placed in food to exert a technological effect. They are likely to be present in larger quantities than additives and should therefore be evaluated. A possible source of such foods is the genetic modification of plants, animals and microorganisms which results in new substances appearing in the food, or the introduction of a new process which leads to a substantial modification of the chemical or nutritive properties of the food. Biotechnology is also being examined from the environmental and agronomic point of view and Council Directives already exists on environmental questions. It is important both to consumers, industry and farmers that simple administrative procedures are followed so that the evaluation of food safety by the Scientific Committee for Food is coordinated with the examination of other aspects by the relevant expert committees.

There is, as yet, no general food contaminant legislation in the Community although specific measures are laid down for some items such as potable and mineral water and radioactive contamination. The Commission's services have collected information on Member States' legislation and a draft directive has been prepared which defines contaminants on the basis of the Codex definition and sets out procedures for laying down common rules. This is a subject which merits a global approach since it is important to evaluate the total tolerable intake from all foods and tolerances cannot be set for one group of foods without considering other sources.

Microbiological contamination has been the source of a number of food scandals in recent years. There are specific measures in place or under consideration for animal products in agricultural policy. However the food chain is complex [6] and rapid developments are taking place in the retailing and catering industries with prepared dishes, microwaveable foods, cook-chill and boil-in-bag products which require a general approach to the question. It would seem desirable to have a framework within which specific rules, codes of practice, and standards could be drawn up. A quality assurance (QA) approach will lead to a much higher degree of security than repressive control. Such an approach is being followed by the USA Food and Drug Administration (FDA) for seafoods where the FDA provides a customer-paid QA auditing service for seafood firms.

The 'perspective' of the design for 1992 has been determined in broad outline and can be seen to be convergent; what is now required is the filling in of the detail. The extensive programme of implementing acts is described in detail by Gray [7]. It is clear that with the current resources available to the Commission the large number of implementing directives will take some time to put in place. In the meantime, trade in foods for which harmonized measures have not yet been laid down will be subject to Article 30 of the EEC Treaty.

In order to clarify the position, the Commission adopted a second Communication of Food Law [8], some important conclusions of which are:

'... As far as the trade description of an imported foodstuff is concerned the importer can in fact choose:

— either to keep the name under which the product is lawfully marketed in the Member State of manufacture,
— or to adopt the trade description under which similar products are marketed in the importing Member State.

Indeed, there is no reason why the product should not bear both trade descriptions at the same time, provided that this does not create confusion for the purchaser.

The only instance in which this choice can be restricted is where the product presented under one or other, or both, of the trade descriptions differs in terms of its composition or manufacture from the goods generally

known under the same descriptions in the Community to such an extent that
it cannot be regarded as belonging to either category . . . .

. . . The importation of a foodstuff containing an additive that is authorized
in the Member State of manufacture, but prohibited in the importing
Member State, must be authorized if:

— the additive does not represent a danger to public health, taking into
   account the findings of international scientific research and the eating
   habits in the importing Member State, and
— the use of this additive meets a genuine need, in particular of a
   technological or economic nature.

To this end, the Member States must institute an authorization procedure that is easily accessible to traders and can be concluded within a
reasonable time (not more than 90 days). Under such a procedure it will be
for the responsible national authorities in the Member States to demonstrate
that the refusal to grant authorization is justified on grounds relating to the
protection of public health. Furthermore, it must be open to traders to
challenge before the courts an unjustified failure to grant authorization.

These principles also apply in cases where the importation of a foodstuff
lawfully produced and marketed in another Member State is restricted on
grounds other than the protection of health . . . .'

Although it would seem that legally the problem of free circulation is resolved by
article 30 there is an urgent need to press ahead in preparing and adopting
implementing measures to resolve uncertainties for the consumer and the industry.
To gain acceptance measures should:

— be based on sound science;
— take a coherent approach to a problem and not be a simple ad hoc amalgam of
   existing measures;
— be formulated with the objective of effective implementation and enforcement.

A recent in-depth review of official chemical and bacteriological testing of food in
Canada [9] has revealed a considerable lack of uniformity in approach to the way in
which priorities are decided. The Canadian government has attempted from this
examination to design a comprehensive plan where testing frequency is related to the
degree of risk presented by a food.

It could be argued that such an approach should not be limited to comparative
studies on enforcement but should be extended to take in the formulation of detailed
implementing rules so that the relationship between a rule, its ease of enforcement
and risk could be developed. In other words an approach should be developed which
allows the effectiveness of enforcement of various solutions to be taken into

consideration when these measures are laid down. The first guiding principle should be that of simplicity. A typical case is that of food additives, where there are currently over 100 000 national provisions in EEC Member States. For sulphite alone there are more than 500 detailed rules, a complex problem for a manufacturer seeking to comply and a controller seeking to enforce. Mollenhauer [10] explains how this admissible daily intake (ADI) of food additives is calculated. He states that, in fixing conditions of use from ADI, the large safety factor, inherent in the ADI calculation means that for food intake data '. . . a lesser accuracy is required than for clinical or other research purposes . . .'.

If this is so, then is it justifiable to lay down complex conditions of use which are difficult to enforce? Surely the auditing of additive intakes, which may change rapidly with the effects of the further integration of the Common Market, is much more important to consumer well-being than an over-detailed set of rules based on scant intake data. The users of food additives have much information on their use and intake which is not available to public authorities, and it would seem reasonable if the authorization of an additive created an obligation on users to continue to provide information on its intake so that its use could be followed over a period of time. In this way the original intake assumptions could be reassessed. It is interesting to note that a ten-year survey of additive intake in Japan based on the analysis of sample diets has shown intake levels to be well below those assumed at the time they were fixed.

If public resources are to be used effectively there is a need for a coherent basis for risk assessment which allows evaluation of risk from various sources to be compared. The Commission's services have agreed to examine this question with the USA FDA within the framework of their cooperation to see whether a common approach to risk assessment can be found.

With the increase in the demand for scientific opinions imposed by the directives already agreed and those areas yet to be covered there is an urgent need to strengthen the infrastructure of the Scientific Committee for Food. Following consultation with the Member States the Commission is preparing a proposal for cooperation in scientific assessment in matters related to food. National bodies involved in this work would cooperate in the preparatory work necessary for the provision of the opinions of the Scientific Committee for Food which are now mandatorily required by a number of Council Directives [11]. The cooperation would not be limited only to these aspects but would deal with any questions of concern in food safety. These could include also joint work in the intake of food additives or the coordination of microbiological surveys.

If sound decisions are to be made on food additives it is essential to have good intake data and this has been referred to previously but the problem of reliable data also exists in the field of microbiology. The recently formed UK Committee on the microbiological safety of food in its report stated that

'. . . although a considerable amount of microbiological surveillance of food does go on it is at present largely *ad hoc* and uncoordinated and therefore of little use in any general and ongoing assessment of microbiological quality. Little of the data can provide any information on the levels of contamination of any particular food with any particular pathogenic organism under any particular conditions of production and distribution or over any period of time .... [12]'.

It is quite unrealistic to formulate or to administer food law if the information on which decisions are being taken is unsatisfactory and surveys of this kind would seem to be a suitable topic for cooperation.

The Council directive on the official control of foodstuffs [13] already provides for the coordination of sampling programmes and this will enable results to be assessed on a comparative basis leading to an increase in effectiveness.

Finally there remain those aspects of food law which are not subject to harmonization. There is a rapidly growing interest in the food sector in the use of QA schemes conforming to the ISO 9000 (EN 29000) series of standards which provide a common basis having international recognition. The recent Commission proposal for a Community scheme for industrial goods [14] could also be applied to the food sector. The Commission will shortly be putting forward a proposal on food quality orientated on a QA approach which can offer a better guarantee to the consumer than mandatory requirements enforced by repressive controls. The resultant simplification would greatly facilitate the task of public control while providing the necessary flexibility for new standards to be developed to keep pace with innovation. A detailed examination of such an approach is to be found in the second Creyssel report [15].

Even when all the 'essential' measures are harmonized, the large number of national 'non-essential' measures will continue to pose many problems for official food controllers, who are not in a position to know, let alone enforce the conformity of a foodstuff to, a national law of the Member State of origin although they will be obliged by Article 30 to accept goods made legally in other Member States. The question could be asked, whether there is not a better way than that of mandatory legal provisions to achieve the ends for which these measures were designed.

Food law should be neutral and should not favour the economic interests of procedures or manufacturers or distort competition on the marketplace. Its main objectives are to render illegal the production and sale of unwholesome food and to ensure that the consumer is not misled by unfair trading practices. To go beyond these objectives is to infringe the basic freedoms which are inherent in our open democratic society and which are the cornerstone of the EEC Treaty. The reassessment of food law which is being carried out in the 1992 process is an ideal opportunity for governments to re-examine national measures which do not serve these objectives, with a view to doing away with those now made redundant by modern food law and replacing by appropriate quality standards those which they deem to be necessary.

The perspective to 1992 seems then to offer a view of a landscape in which consumer protection is assured and protected by effective control and competition to improve food quality may take place within a system which assures fair trading. With the recent developments in Central Europe and the negotiation of the European Economic Area, EC food law is likely to have a much wider impact than was envisaged when the present policies were conceived. Belgium, as a founder member of the Community, can be proud of making an outstanding contribution to these developments.


## REFERENCES

[1] *Loaves and Fishes, an illustrated history of the Ministry of Agriculture, Fisheries and Food.* MAFF, London.

[2] Compte-rendu de la quatrième Conférence internationale de la Chimie (1923) pp. 133–146.

[3] Case No. 12/74 (Sekt) judgement given on 20.2.1975, reported in *European Court Reports 1975*, p. 181.

[4] Case No. 120/79 (Cassis de Dijon) judgement given on 2.2.1972, reported in *European Court Reports 1979*, p. 649.

[5] Completion of the Internal Market Community Legislation on Foodstuffs. COM(85)603, CEC, Brussels, Belgium.

[6] Gray, O. W. Pressure Groups and their Influence on Agricultural Policy and its Reform in the European Community. PhD Thesis, University of Bath, 1990.

[7] Gray, P. S. Food law and the internal market: taking stock. *Food Policy*, **15** No. 2, 111–121.

[8] Communication on the free movement of foodstuffs within the Community. OJ C 271, 24 October 1989, p. 3.

[9] Consolidated strategy for the assessment of chemical and biological hazards in agri-food commodities, 1989–1993. Food production and inspection branch, Agricultural Canada, Ottawa.

[10] Mollenauer, H. Control of food additives. *Food Control* **1** No. 2, 69.

[11] Council Directives on extraction solvents, flavouring agents food additives, materials and particles in contact with foodstuffs and on foods for particular nutritional use.

[12] Richmond, M. *et al.* (1990) *The report of the Committee on the Microbiological Safety of Food (part 1) to the UK Government.* HMSO Dd292322C, London.

[13] Council directive on the official inspection of foodstuffs. OJ L 186, 30 June 1989, p. 23.

[14] A global approach for certification and testing, quality measures for industrial products. CAM(89)209, 24 July 1989, CEC, Brussels.

[15] Creyssel, P. La certification des systèmes d'assurance qualité dans le secteur agro-alimentaire. Report to the French Minister of Agriculture, Paris, September 1990.

# I.2

# Food policy — implications for the nutritional sciences

**Michael J. Gibney**
Division of Nutritional Sciences, Department of Medicine, Trinity College
Medical School, St. James Hospital, Dublin 8, Ireland

Food policy has many dimensions, of which five can be given prominence (Fig. 1).
Food policy must take account of trade issues, environmental issues, agricultural
issues, strategic issues and of course health issues. None of these should be

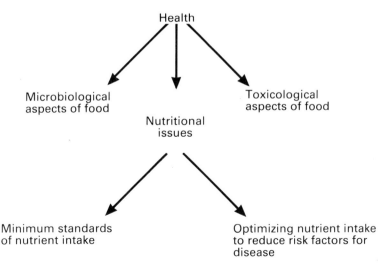

Strategic      – The food supply in emergencies
Agricultural   – The Common Agricultural Policy
Trade          – Free trade effects on world commodity prices
Environmental – Organic and integrated agriculture

Health

Microbiological
aspects of food

Nutritional
issues

Toxicological
aspects of food

Minimum standards
of nutrient intake

Optimizing nutrient intake
to reduce risk factors for
disease

Fig. 1 — The components of food policy.

considered in isolation from the others. The health aspect of food policy has three main elements: toxicology, microbiology and nutrition. Thus the terms food policy and nutrition policy are not interchangeable, for the latter is only one sub-component of the former. The nutrition component of food policy is itself comprised of two elements. The first centres around the need to direct food consumption such that minimum standards of intake, the recommended dietary allowances, are achieved. This often involves policy decisions in the area of food fortification, social welfare schemes, school lunch programmes and such like. The second element of nutrition policy involves the direction of food consumption to achieve a balance of nutrient intake commensurate with optimum health. It is this latter aspect which this paper will address. It does not set out to challenge the need for such. Rather it sets out to look at the issues for nutritional science arising from such aspects of food policy.

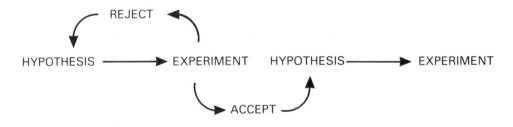

Fig. 2 — The scientific method.

## SCIENCE AND POLICY

The knowledge of man is two kinds, the subjective and the objective. The former is characterized by our knowledge of history, theology, economics, literature and so on. The latter is dominated by science, where our knowledge is objective on the basis of its origins in experimentation. Science proceeds as shown in Fig. 2. From an hypothesis an experiment is designed to test its validity, in effect to refute the hypothesis. If the outcome of the study indicates rejection of the hypthesis, the scientific team must re-think and re-formulate their hypothesis. If the outcome of the experiment indicates acceptance of the hypothesis, it will remain so until it is re-designed and re-tested to take our knowledge further. The essential feature of science is its change of views in the light of new experimental evidence, such as our view of the physical world which has progressed from Copernicus to Newton, to Einstein and the big-bang theory of today.

This scientific process has two important lessons to those charged with the formulation of nutrition policy. The first is that science changes its mind. This can best be illustrated from the nutrition literature. In the late 1950s and early 1960s, two groups, one in Harvard and one in Minnesota, worked to establish the quantitative relationship between changes in the intakes of dietary fats and changes in blood cholesterol. They both concluded that polyunsaturated fatty acids lower blood cholesterol, that saturated fatty acids raise blood cholesterol and that in this respect monounsaturated fatty acids are neutral (Keys *et al.* 1957, Hegsted *et al.* 1965). That policy, of increasing polyunsaturated fatty acid intake at the expense of saturated fatty acid intake has dominated the events in this field for the last two decades and still continues to be the dominant policy with regard to dietary fat. Since 1985, a number of papers have appeared which provide powerful evidence that monounsaturated fatty acids are as effective as polyunsaturated fatty acids in lowering blood cholesterol without any of the potential side effects of the latter. The relevant literature has been reviewed by several authors (Grundy & Denke 1990). In the light of our understanding of the mechanism that science changes its mind, and then only reluctantly as is its proper conservative duty, we should use scientific data in the formulation of nutrition policy with a degree of humility, for our knowledge of the issues at hand is built on shifting sands.

A second lesson we must learn from an examination of the scientific method is that it generates only a limited number of pieces in the complex jigsaw puzzle relating diet to health. Each piece of the jigsaw, that is each scientific paper, may in itself be verifiable by experimentation. Nonetheless the jigsaw puzzle remains incomplete. It is at this stage that governments, in attempting to formulate a nutrition policy, call upon committees of experts to attempt to resolve the puzzle to the best of their ability. Necessarily, the data they deal with is inconsistent and incomplete. Were it otherwise there would be no need for committees of experts. The sun rises in the east and water boils at 100°C, two facts which do not require an expert opinion. The committee of experts, in evaluating this incomplete and inconsistent data, issue an opinion based on 'the probability of evidence'. That being the case, it is essential that attention be paid to how such committees work, for if this part of the process is in error, all else will subsequently be likewise.

## COMMITTEES OF EXPERTS

The composition of an expert committee may determine its outcome. A committee of vegans is unlikely to extol the virtues of animal foods. Likewise a committee of like-minded nutritionists, adherent to one lobby of opinion, will issue opinions compatible with their common view. Given that expert committees are convened only for the resolution of contentious and difficult issues, it is vital that all opinions are represented. Following 20 years of service to the United Nations–World Health Organization expert committee on protein and energy requirements, Scrimshaw (1976) highlighted the many pitfalls of expert committees, including inadequate first-hand knowledge of the literature, inadequate time to consult documents and entrenched personal opinions. With respect to the need for wide-ranging representation on such committees, he wrote:

> We need constantly to remind ourselves that neither individuals nor com-
> mitttees are always infallible, and that all scientific issues must be addressed

with some humility. In particular, the democratic approach to scientific truth is a contradiction in terms — the truth must often be with the majority, but it is not always. The dissenters should be listened to carefully.

**Table 1** — Recent reports on diet and health. A comparison of possible scales of approach

|  | US[a] | WHO[b] |
|---|---|---|
| Pages | 727 | 150 |
| Topics | 18 | 12 |
| Editors | 10 | 3 |
| Contributors | 69 | 0 |
| Review stages | 6 | 0 |
| Reviewers | 171 | 0 |

[a]DHHS (1988).
[b]WHO (1988).

Table 1 draws some comparisons between two recent reports on diet and health with marked differences in their approach to ensuring that all views are listened to and accommodated. Clearly the more rigorous US approach to this task is more expensive in terms of time, money and administrative support. However, it does set the standard in the management of nutrition policy considerations. Where multi-state organizations such as the World Health Organization or the Commission of the European Communities are responsible for drafting such policy, adequate inter-state and intra-state discussion, embracing all opinions and all sectoral interests, must be carried out. If the expert opinion is flawed, all else will be flawed. If the expert opinion has been truly consultative, all sectoral interest can then work together in confidence.

## NUTRIENT GOALS — QUANTITATIVE OR QUALITATIVE

Expert committees on nutrition are expected to issue guidelines on how the pattern of nutrient consumption should be influenced. They may be either qualitative or quantitative. Qualitative guidelines use terms such as 'increase intake of ...', 'ensure adequate intakes of ...', 'avoid excessive intakes of ...' and 'reduce intakes of ...'. There is no attempt to quantify what is 'adequate' or 'excessive' or by how much consumption should be 'increased' or 'decreased'. There are advantages in this approach which has been consistently favoured by US Federal authorities (Miller & Stephenson 1985). Firstly, such committees do not have to justify figures. Secondly, by avoiding nominating their targets with any degree of precision, they have latitude to be content with modest changes or be unhappy with subsequent substantial changes in food intake where evidence suggests that further change is needed.

Quantitative guidelines, such as 'increase fibre intake to 25 g/d' or 'reduce fat intake to 35% of dietary energy' have been favoured in Europe (WHO 1988). They are clearly preferred by planners, who can project what the effects of a given change in the food supply would have on the intake of a given nutrient. Their disadvantage is that a figure has been declared and either it's right or it's not.

The committee should give detailed reasons as to precisely why that figure was picked. It implies a detailed knowledge, on the one hand, of the dose–response relationship between a nutrient and a physiological parameter (fat and blood cholesterol or fibre and faecal output) and on the other hand of the dose–response relationship between the appropriate physiological parameter and the disease in question (blood cholesterol and heart disease or faecal output and diseases of the gastrointestinal tract). Regrettably, significant detail as to the basis of the choice of the figure in question is rarely given, largely because the nature of these dose–response relationships is unknown. Committees on toxicology, who formulate acceptable daily intakes of food additives, of trivial public health significance, do justify their figures. Nutrition committees should take a leaf from the reports of toxicology committees. One problem, as yet unresolved, is how fat intake should be expressed relative to energy intake.

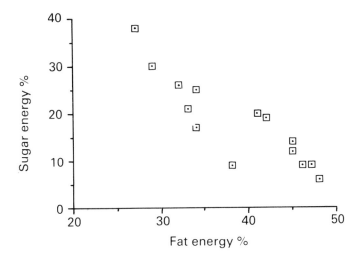

Fig. 3 — The inverse relationship between the percentages of dietary energy from sugar and from fat. Based on four recent surveys in selected northern EC states (Tables 3–6; Gibney 1990).

## FAT–ENERGY INTER-RELATIONSHIPS
Fat intake is usually expressed, not in absolute terms, but in terms of its contribution to energy intake. Several studies have shown that a clear inverse relationship exists between the percentage of energy from fat and that from sugar (Fig. 3; Gibney 1990) and alcohol (Fig. 4; Gibney et al. 1989). The percentage of energy from fat is normally distributed (Bingham et al. 1981) while that for sugar and alcohol and most foodstuffs is not (Haralsdottir et al. 1987), tending to be heavily skewed toward the

lower end of intakes. Fat energy intakes can thus change considerably without any absolute change in fat intake, simply because the energy derived from alcohol and sugar changes. The data from the Seven Countries Study is widely used to determine the optimum level of fat in the diet but no consideration is ever given to the potential confounding effects of both sugar and alcohol in the interpretation of this data. If one looks at Fig. 3 then one finds no subgroup which achieves both the WHO targets of fat and sugar (30% fat energy, 10% sugar energy). Should nutritionists begin to consider expressing fat intake as a percentage of non-sugar, non-alcohol energy to remove these powerfully confounding effects?

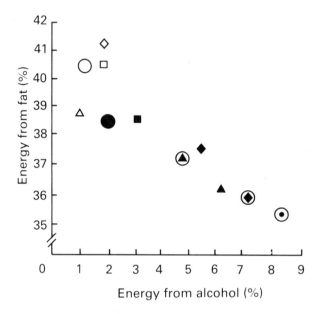

Fig. 4 — The inverse relationship between the percentages of dietary energy from alcohol and from fat. Based on seven recent studies in the UK and Ireland (Gibney *et al.* 1989).

The purpose of this paper is to explore the issues for nutritional science in the health aspects of food policy. Of all such issues which can be raised, this see-saw relationship between energy from fat and energy from sugar and alcohol, is the most important. We cannot begin to interpret international patterns of energy derived from fat or time-related trends in the energy derived from fat with any degree of confidence until these issues are resolved. In the period 1980 to 1986, fat intake in the UK fell by 6 g/d. However, energy intakes also fell because sugar and white bread intakes fell (Buss 1988). Thus the percentage of energy from fat actually increased. Was this a success or a failure?

## FORMULATING PRACTICAL DIETARY ADVICE

If the objective is to reduce our intake of fat then some mechanism must be constructed to ascertain the pattern of food intake which leads to an unacceptably high level of fat intake. The traditional approach has been to identify the sources of fat in the national diet and then to target those sources. In general, meat, spreadable fats and dairy products account for 70% or more of fat intake in European countries. Not surprisingly therefore, these have borne the brunt of criticism of the quality of our diets. However, if within a population, subgroups with contrasting intakes of fat (% dietary energy) are examined then little difference is found in the sources of fat between groups with high- or low-fat diets. This is shown in Table 2 for four northern EC states for which data exists. Therefore, knowing the sources of fat in the diets of subgroups does not necessarily help us construct a mechanism for identifying the pattern of food intake which predisposes to a high-fat diet.

**Table 2** — Sources of fat (%) in groups consuming high or low levels of dietary fat in four European states

| Dietary fat: classification | Denmark[a] | | Ireland[b] | | UK[c] | | Netherlands[d] | |
|---|---|---|---|---|---|---|---|---|
| | Low | High | Low | High | Low | High | Low | High |
| Meat | 219 | 21 | 18 | 23 | 21 | 21 | 20 | 19 |
| Spreadable fats | 56 | 48 | 32 | 25 | 20 | 17 | 28 | 33 |
| Milk | 7 | 9 | 14 | 9 | 12 | 10 | 12 | 11 |
| Biscuit and cakes | — | — | 10 | 12 | 11 | 15 | 5 | 4 |
| Cheese | 8 | 9 | — | — | 8 | 6 | 8 | 9 |
| Eggs | 3 | 4 | 5 | 5 | 5 | 5 | 2 | 2 |
| Chips | — | — | — | — | 5 | 4 | — | — |

[a]Haraldsdottir *et al.* (1987).
[b]M. Gibney (unpub.)
[c]MAFF (1983).
[d]WVC (1988).

An alternative approach is to define two subpopulations, at say the upper and lower quartiles of fat intake as a percentage of energy, and then ask the questions: what do people eat to get it right and what do people eat to get it wrong? Tables 3 to 7 give details of food intake in groups with low- and high-fat diets from five recent studies which have permitted such comparisons to be made. The data does not present a consistent pattern. In Denmark, for example, it does not appear as though meat or milk distinguishes high and low fat eaters while butter does (Table 4). In low socio-economic groups in Ireland the same could be said about butter and about meat (Table 6). Among young Dutch adults a clear pattern emerges with the high-fat group having higher intakes of meat, spreadable fats and dairy products compared to the low-fat group.

However, these groups are defined on the basis of the percentage energy from fat and that, as has been seen in the previous section, is determined not only by fat intake but by the intakes of alcohol and sugar as is evident in the data for young Dutch adults in the case of alcohol and of Irish low socio-economic groups in the case of sugar (Tables 7 and 6 respectively).

**Table 3** — Patterns of food intake in adult Irish of mixed socio-economic background, classified into low- (<35%) or high- (>40%) fat energy intakes

|                        | Men     |      | Women   |      |
|------------------------|---------|------|---------|------|
| Fat (% energy)         | <35     | >35  | <35     | >35  |
| (no. in study)         | (8)     | (11) | (6)     | (12) |
| Energy (MJ/d)          | 11.4    | 11.6 | 7.0     | 9.0  |
| % fat energy           | 33      | 45   | 34      | 45   |
| Fibre (g/10MJ/d)       | 21      | 20   | 33      | 24   |
| Food intakes (g/10MJ/d) |        |      |         |      |
|   Milk       | 226     | 312  | 274     | 234  |
|   Butter     | 14      | 29   | 9       | 24   |
|   Margarine  | 20      | 10   | 34      | 12   |
|   Spreadable fats | 34 | 39   | 43      | 36   |
|   Meat       | 147     | 175  | 120     | 151  |
|   Fruit      | 49      | 79   | 91      | 91   |
|   Bread      | 204     | 164  | 273     | 164  |
|   Potatoes   | 345     | 187  | 249     | 164  |
|   Other vegetables | 162 | 166 | 219    | 147  |
|   Sugar      | 68      | 10   | 19      | 22   |
|   Sweets and chocolates | 15 | 6 | 10   | 10   |
|   Biscuits and cakes | 41 | 60  | 29     | 69   |

*Source*: Gibney *et al*. (1989).

These data resolve one issue. They show the weakness in the strategy of looking at the sources of fat in the national diet and basing the strategy on that finding. They also show the need for more imaginative processing of national dietary survey data to establish markers of good and bad eating practices.

## SETTING GOALS FOR NUTRIENT INTAKE

When quantitative dietary guidelines are set, they are inevitably centred around the mean. Typically it would be proposed that intakes of fat be reduced from around 40% of dietary energy to about 35% of dietary energy. In considering why the mean is used and not the distribution about the mean, a number of reasons emerge. To begin with, that is how it is always done and science tends to be conservative. Secondly, population data of fat-intake–risk-factor relationships usually involve mean values of fat intake for groups with varying degrees of risk-factor prevalence. Thirdly, and perhaps most importantly, data on distribution about the mean is usually very limited.

In estimating risk-assessment exposure to food additives or contaminants, mean values are never used. The 90th percentile of intake is the value of choice (National

**Table 4** — Patterns of food intake in adult Danes classified into quartiles of fat intake

|  | Fat energy quartiles | |
|  | Highest | All others |
| --- | --- | --- |
| (n=) | (226) | (2016) |
| % fat energy | 50 | 40 |
| Fibre (g/10MJ/d) | 19 | 23 |
| Food intake (g/10MJ/d) |  |  |
| Full milk | 292 | 405 |
| Low-fat milk | 136 | 108 |
| Skimmed milk | 85 | 145 |
| Cheese | 48 | 44 |
| Meat | 116 | 104 |
| Fish | 24 | 24 |
| Eggs | 32 | 33 |
| Butter | 31 | 17 |
| Margarine | 37 | 33 |
| Cereal products | 180 | 207 |
| Table sugar | 18 | 24 |
| Fruit | 97 | 168 |
| Potatoes | 114 | 128 |
| Other vegetables | 109 | 132 |

*Source*: Haraldsottir *et al.* (1987).

Academy of Sciences 1981). It is perhaps time that nutrient goals also accept the complexity of distributions of intakes and move from mean-orientated targets to distribution-orientated targets. The preceding section has argued for a contrast to be made between the pattern of foods consumed by individuals at the upper and lower quartiles of intake of a targeted nutrient. The upper quartile of fat intake in Denmark is 50% energy (Table 4). An alternative to looking at mean intakes would be to seek a reduction in the number of people exceeding that figure, or to reduce the mean intake of those individuals in the upper quartile of intake. This approach may be more rewarding from the health education point of view since substantial changes could be made at the extreme of intakes with only minimal impact on the mean intake. Relatively simple indices of high-fat eating, the need for which has been argued in the previous section, could be used to monitor response to the nutrition component of health education programmes.

## END-POINTS OF NUTRITION INTERVENTION PROGRAMMES

It is not infrequent to hear epidemiological nutritionists discuss the diet of different populations which have markedly different mortalities from coronary heart disease.

**Table 5** — Patterns of food intake in adult English classified into both low-fat (<35% energy) and above-average fibre (>20 g/d) intakes, or high-fat (>35% energy) and below-average fibre diets

|  | Men | | Women | |
|---|---|---|---|---|
| Fat (% energy) | <35 | >35 | <35 | >35 |
| Fibre (g/d) | >20 | <20 | >20 | <20 |
| (n=) | (7) | (83) | (3) | (99) |
| Energy (MJ/d) | 12.8 | 11.8 | 8.7 | 8.3 |
| % fat energy | 32 | 41 | 34 | 41 |
| Fibre (g/10MJ/d) | 22 | 18 | 28 | 19 |
| Food intakes (g/10MJ/d) | | | | |
|   Milk | 237 | 257 | 311 | 340 |
|   Butter | 9 | 14 | 16 | 13 |
|   Margarine | 13 | 9 | 8 | 10 |
|   Spreadable fats | 21 | 23 | 24 | 23 |
|   Meat | 120 | 146 | 136 | 133 |
|   Fruit | 41 | 51 | 69 | 67 |
|   Bread | 96 | 117 | 97 | 101 |
|   Potatoes | 198 | 109 | 95 | 98 |
|   Chips | 34 | 41 | 38 | 35 |
|   Baked Beans | 63 | 17 | 116 | 16 |
|   Other vegetables | 137 | 108 | 130 | 124 |
|   Sugar | 87 | 45 | 46 | 25 |
|   Biscuits and cakes | 113 | 122 | 125 | 142 |
|   Sweets | 3 | 6 | 7 | 6 |

*Source*: Nelson (1985).

Equally it is not unusual to hear epidemiological nutritionists discuss time-related changes in mortality with dietary trends. Such inferences are lacking in serious scholarship. We recognize the multiplicity of variable risk factors for coronary heart disease (smoking, hyperlipidaemia, hypertension, diabetes, obesity etc). We must also recognize the contribution which is made to coronary heart disease by a new generation of effective drugs, by advanced diagnostic techniques, by more effective screening and by advances in acute and chronic clinical managment of the disease. Within this, we must not attempt, without adequate data, to isolate the effects of diet. Rather, nutritionists must insist that the end-points of nutrition intervention are expressed in nutritional terms of intakes of foods and nutrients. If a campaign to lower fat intake is successful, its success is measured in the mean intakes or the distribution of intakes of fat. It must not be measured in terms of the risk factor which it affects since many other factors could alter that risk-factor prevalence. Nor should it be measured in terms of mortality. Public health programmes to modify lifestyle

**Table 6** — Patterns of food intake in adult Irish of low socio-economic background, classified into upper or lower quartiles of fat intake

|  | Men | | Women | |
|---|---|---|---|---|
| (n=) | (7) | (83) | (3) | (99) |
| Energy (MJ/d)[a] | 14.0 | 11.7 | 10.8 | 7.8 |
| % fat energy | 29 | 47 | 27 | 46 |
| Fibre (g/10MJ/d) | 16 | 19 | 13 | 17 |
| Food intakes (g/10MJ/d) |  |  |  |  |
| Milk | 393 | 385 | 500 | 679 |
| Butter | 24 | 48 | 18 | 44 |
| Margarine | 2 | 3 | 4 | 1 |
| Spreadable fats | 26 | 51 | 22 | 55 |
| Meat | 84 | 128 | 71 | 117 |
| Fruit | 17 | 15 | 25 | 43 |
| Bread | 176 | 166 | 149 | 190 |
| Potatoes | 192 | 201 | 159 | 150 |
| Chips | 41 | 31 | 36 | 26 |
| Other vegetables | 108 | 145 | 113 | 105 |
| Sugar | 144 | 15 | 196 | 9 |
| Biscuits and cakes | 34 | 43 | 19 | 33 |
| Sweets | 11 | 15 | 15 | 23 |

[a]Excludes alcoholic beverages.

*Source*: Lee & Gibney (1988).

are multi-faceted. Each facet must have its own end-point while the end-point of the overall programme will of course remain the prevalence of the disease in question.

## THE SIDE-EFFECTS OF NUTRITION INTERVENTION

Public health programmes involving nutritional aspects of disease prevention often focus on a limited number of nutrients — fat, fibre, salt, sugar and alcohol. However, the totality of food intake must not only satisfy the end-points of such nutrition intervention programmes but must also satisfy the full range of the daily nutrient requirements. The two should never be dissociated. Consider the data for Irish females of low socio-economic status (Table 6). Those with a higher-fat diet eat more meat. Let us imagine a campaign was initiated to reduce fat intake in this group which included limitations on red meat intake. Since their prevailing iron intakes average only 7.7 mg/d, such a campaign would make an already undesirable level of iron intake unacceptably low. The implications of nutrition intervention for overall nutrient intake must always be borne in mind. One could quite readily imagine other hypothetical problems. A campaign to reduce sugar intake could in some women

**Table 7** — Patterns of food intake among Dutch 18-year-olds with fat energy intakes
<35, 35–40 and >40% of dietary energy

|  | Fat energy group | | |
|  | <35% | 35–40% | >40% |
| --- | --- | --- | --- |
| Fat (% energy) | 31.5 | 37.6 | 43.8 |
| (n=) | 53 | 72 | 60 |
| Energy (MJ/day) | 14.5 | 14.9 | 13.9 |
| Fibre (g/10 MJ/day) | 24 | 23 | 22 |
| Food intake (g/10 MJ/day) | | | |
|   Milk, full-fat | 186 | 233 | 270 |
|   Milk, half-fat | 94 | 89 | 78 |
|   Milk, skimmed | 94 | 40 | 47 |
|   Cheese | 21 | 27 | 32 |
|   Spreadable fats | 17 | 27 | 38 |
|   Meat and meat products | 77 | 82 | 92 |
|   Fruit | 119 | 136 | 98 |
|   Bread | 172 | 167 | 176 |
|   Potatoes | 202 | 187 | 192 |
|   Vegetables | 105 | 98 | 96 |
|   Sugar, sweets and sweet spreads | 62 | 55 | 44 |
|   Alcoholic drinks | 414 | 240 | 139 |

*Source*: Hulshof & Ockhuizen (1988).

lead to levels of energy intake which, even in the short term, would not be acceptable in pregnancy. Raising awareness among adolescent females that milk is fattening could jeopardize calcium intakes. Careless, ill-conceived and naive nutrition education programmes should not be tolerated by the science of nutrition.

## SELLING THE MESSAGE

It is one thing to formulate a correct message. It is quite another thing to make people listen. Experience with tobacco, alcohol, drugs, AIDS and other such issues indicates that knowledge alone is not enough to motivate people. Many people will quite simply not believe the message, citing acquaintances who obeyed all the rules and died young and others who disobeyed all the rules of a healthy lifestyle and who lived long lives. The collective wisdom of decades of international collaborative study has been required to establish links between diet and risk factors. Therefore we should not expect people to ignore their own life-experience in favour of complex epidemiological data. Even if people believe the message they may be indifferent to it for one reason or another. Equally they may be interested in the message but find it difficult to put into practice. There are many possible negative attitudes to health education messages including nutrition education. Therefore it behoves nutritionists to begin to build bridges with the behavioural scientists to try and understand the forces which both stimulate and inhibit changing attitudes to diet. If we ignore this serious deficit in our understanding of the behavioural forces which influence food choices, we run the risk of having large tracts of the population ignoring the message, just as has happened for tobacco, smoking, drugs, AIDS etc. Most of these health education programmes have, at the end of the day, culminated in legislative intervention. Will that happen for nutrition?

## CONCLUSIONS

The relationship between diet and risk factors is always arguable. Notwithstanding that, dietary intervention is rightly seen to play its role in public health programmes. However, there are serious issues which nutritional science must address if the nutrition component of these health education programmes is to be successful. It is therefore essential that the locus for the solution to many of these issues be placed within nutritional science and not within community health or epidemiology, where the complexity of nutrient–nutrient, nutrient–food and food–food interrelationships is often ignored.

## REFERENCES

Bingham, S., McNeil, N. I. & Cummings, J. H. (1981) The diet of individuals: A study of a randomly chosen cross-section of British adults in a Cambridgeshire village. *British Journal of Nutrition* **45** 23–25.

Buss, H. (1988) Is the British diet improving? Proceedings of the Nutrition Society **47** 295–306.

Cunningham, K. & Lee, P. (1990) Irish National Diet Survey. The Institute of Nutrition & Dietetics in Ireland, Dublin.

DHSS (1988) *The Surgeon General's Report on Nutrition and Health*. DHHS (PHS) Publication No. 88-50210.

Gibney, M. J., Moloney, M. & Shelley, E. (1989) The Kilkenny Health Project: food and nutrient intakes in randomly selected healthy adults. *British Journal of Nutrition* **61** 129–137.

Gibney, M. J. (1990) Dietary guidelines: a critical appraisal. *Journal of Human Nutrition & Dietetics* **3** 245–254.

Grundy, S. M. & Denke, M. A. (1990) Dietary influences on serum lipids and lipoproteins. *Journal of Lipid Research* **31** 1149–1172.

Haraldsdottir, J., Holm, L., Jansen, J. & Moller, A. (1987) *Danskernes Kostvaner 1985* 2. Hvem spiser Lard? Levensmiddelstyrelsen Publication Nr 154, Soborg.

Hegsted, D. M., McGandy, R. B., Myers, M. L. & Stare, F. J. (1965) Quantitative effects of dietary fat on serum cholesterol in man. *American Journal of Clinical Nutrition* **17** 281–295.

Hulshof, K. F. A. M. & Ockhuizen, T. (1988) TNO — rapport number v 88, 451. Consumptiepatroon van 18 — jarige mannen; met speciale aandacht voor de vetconsumptie. CIVO–Instituten TNO, Zeist.

Keys, A. A., Anderson, J. T. & Grande, F. (1957) Predictions of serum cholesterol responses of man to changes in fats in the diet. *Lancet* ii, 959–964.

Lee, P. & Gibney, M. J. (1988) Patterns of food and nutrient intake in a Dublin suburb with chronically high unemployment. Combat Poverty Agency, Dublin.

Miller, S. A. & Stephenson, M. G. (1985). Scientific and public health rationale for the dietary guidelines of Americans. *American Journal of Clinical Nutrition* **42** 739–745.

National Academy of Sciences (1981) *Assessing Changing Food Consumption Patterns,* by the Committee on Food Consumption Patterns, Food & Nutrition Board, National Research Council. National Academy Press, Washington DC.

Nelson, M. (1985) Nutritional goals from COMA and NACNE: How can they be achieved? *Human Nutrition Applied Nutrition* **39A** 456–464.

Scrimshaw, N. S. (1976) Strengths and weaknesses of the committee approach. An analysis of past and present recommended dietary allowances for protein and health in disease. *New England Journal of Medicine* **294** 198–203.

WHO (1988) *Healthy Nutrition. Preventing nutrition-related diseases in Europe.* World Health Organization Regional Publication Series No 24, WHO, Geneva.

Ministerie van Welzijn, Volksgezondheid en Cultuur (WVC) (1988) *Wat Eet Nederland. Resultaten van de voedselconsumptiepeiling 1987–1988.* DOP, Den Haag.

# I.3

## An industry's point of view

**Irina du Bois**
Nestlé, Vevey, Switzerland

It is impossible to talk about the food industry without paying tribute to the major discoveries in modern food technology in the nineteenth and twentieth centuries. The father of food industry is without doubt the Frenchman, Nicolas Appert, the inventor of canned foods in 1810 — bottled conserves first, followed rapidly by metal cans.

In the history of conservation, *milk*, of course, holds a privileged position because of its importance in human nutrition as well as due to the unique technological developments linked to its transformation; the first sweetened condensed milk factory was opened in 1856 in the USA, followed 10 years later by the début in Cham of the Swiss sweetened condensed milk industry. The thermic treatment invented by Louis Pasteur in 1865 for wine, was later applied to milk — the famous pasteurization — and it was only a century later, in 1961, that the extraordinary combination of two technologies, one for the product and one for the packaging, permitted the sale of 'uperized' milk.

A special mention must go to *infant foods*, of which the father — or should we say the mother? — was without doubt Henri Nestlé, with his infant milk cereal, back in 1866 in Vevey.

After the milk, it was *chocolate*. From 1819 onwards in Switzerland, Cailler, Suchard, Kohler and Lindt made our mouths water with the industrial production of chocolate, and in 1875 Daniel Peter achieved the marriage of the century, milk chocolate.

Towards the end of the 1860s, the invention of *margarine* by the Frenchman Hyppolyte Mège-Mouriès was to profoundly and durably modify food habits.

Two of the most important ingredients to culinary preparations must also be mentioned: the *meat extract* discovered by Max Pettenkofer in 1847 which was analysed and signed later by Justus Liebig, and the all vegetable *Maggi seasoning*, obtained in 1866 by Julius Maggi.

If the era of *refrigeration* began in 1835 with the patent by the American Perkins, we had to wait until 1920 to have individual refrigerators in American homes, and 1930 to have frozen foods at our disposal, thanks to the Englishman, Birdseye.

The last major discovery before the Second World War was *soluble* or *instant coffee*, which became available in 1938 developed by the Morgenthaler team in Nestlés laboratories.

After the Second World War, in the 1950s, ultrafiltration, reverse osmosis and electrodialysis for the milk industry should be mentioned, as well as irradiation in the 1960s — and then in 1970 the discovery of isomerosis, the high fructose glucose syrup. Finally, let us mention the fantastic progress that plastic packaging has made over the past 40 years.

And *food legislation*? What has been its evolution, its history? All food legislation is aimed at:

— the protection of public health
— the protection against fraud
— the assurance of fair trade practices
— information of the consumer.

During the Middle Ages, the modern era and until the end of the nineteenth century, the repression of fraud was the major problem of the authorities. Not only the authorities, but also the corporations have intervened over the years to ensure the protection of quality, the authenticity and the good repute of products. At that time, the control of foodstuffs was limited to organoleptic tests (aspect, odour, taste). It was only in the middle of the last century, with the knowledge of chemistry, that more efficient controls became possible. From that time, the first national laws destined for the protection of health and against fraud were published in various countries: in Great Britain (1860), in Germany (1879), in Belgium (1890), where we are today celebrating the hundredth anniversary, in France and in Switzerland (1905), and in the USA (1938).

All these laws in reality have followed the objectives mentioned above. However, due to historic differences, differences in mentalities, different food habits, they were translated into practice by a wide variety of diversified regulations. The result was that the food industry — and particularly those present internationally — were confronted with numerous non-tariff trade barriers.

Amongst the various organisms which, over the years, have dealt with these hindrances by endeavouring to eliminate them, the importance of the *Common Market* need no longer be underlined. It is also important to mention the *Codex Alimentarius*, i.e. the Joint Programme on Food Standardization of the United Nations Food and Agriculture Organization and the World Health Organization. The principle of the harmonization of national regulations which governed the policy of the Community between 1957 and 1985, date of the 'new approach' to food law, has already been mentioned. During this period, the food industry — or at least certain industries — made every effort to achieve this harmonization. Through the national trade associations and in the framework of the CIAA, the Confederation of the Food and Drink Industries of the EEC, the industry has used its knowledge, its experience and its ideas towards the progress of Community regulation.

At the same time, in the non-harmonized domains, we have continued to live with very diverse national regulations and, even for certain so-called harmonized domains, we have been obliged to accept the different derogations allowed. In this context, one cannot omit to mention as an example the case of the *EEC Directive on Cocoa and Chocolate Products* which, at the same time, clearly shows the diversity of tastes and gastronomic cultures in Europe and also the regulatory deadlock which we have ended up with. The Directive, in reality, represents a restriction to the free movement of chocolate products throughout the Community.

But let us take a closer look at this. The EEC Cocoa and Chocolate Products Directive was adopted in 1973, at the time when vertical directives were considered to be the key to freedom of trade between EC countries. 1973 was shortly after the accession of the UK, Ireland and Denmark to the EC, which led to the inclusion in the Directive of a number of derogations which acknowledged, on a temporary basis, the existence of different laws and usages in those three countries compared with the original six members. In particular, concerning the critical area of the use of vegetable fats other than cocoa butter, UK, Ireland and Denmark allowed their use in chocolate whereas the others did not. Therefore, due to the impossibility of reaching an agreement, the 1973 Directive forbade the use of vegetable fats in general and authorized derogations to the three newcomers. However, knowing well that they were on dangerous ground, the Directive prudently envisaged (Article 14.2.a.) the possibility of extending the use of vegetable fats to other Member States by saying:

'At the end of a period of 3 years from the notification of the Directive, the Council shall decide on a proposal from the Commission on the possibilities and forms of extending the use of these fats to the whole of the Community.'

The Commission indeed attempted to introduce a new Directive, authorizing the use of vegetable fats in chocolate throughout the Community, but failed in 1986 due to the opposition of the European Parliament.

Thus, since 1973, each Member State has continued to live and eat chocolate as they prefer. What is the situation today? Based on the Directive, chocolate made and sold in the UK, Ireland and Denmark may contain vegetable fats to a maximum of 5% of the total weight of the chocolate, but may **not** be sold under the designation 'chocolate' in the other Member States. Thus free circulation is blocked, a situation which is totally contrary to the actual philosophy of the new approach, to the principles established by the jurisprudence of the European Court of Justice and to the *Communication on the free movement of foodstuffs within the Community* published in 1989. What do the principal actors, the Commission, the Member States and the Industry, have to say? The Commission refers to Article 14.2.a. mentioned earlier and says 'as the three-year limit has been expired for a number of years — this was in 1973 — the question of the use of vegetable fats is no longer covered by the Directive, but must be regulated by the Member States'. The Commission considers thus that chocolate from the UK, Ireland and Denmark may circulate freely and encourages the other Member States to modify their regulations to allow, on their territory, the manufacture of chocolate with vegetable fats.

These same Member States respond, 'Our regulations which forbid the designation "chocolate" both for a product manufactured with vegetable fats locally and imported are based on the 1973 Directive. So, change the Directive and we will adapt our national regulations.' Thus we have a perfect 'tennis match' where we, the food industry, bear the cost. Could the arbitrator of this match be the European Court of Justice? It should be remembered that all judgements taken in the Court of Justice which have paved the way to the free circulation of foodstuffs have been made in the absence of a vertical directive. Therefore, the mere existence of the vertical Chocolate Products Directive means that a 'Cassis de Dijon' type of case might be defeated in the European Court of Justice. But even if successful, it would merely clear the way for British/Irish/Danish chocolate with vegetable fats to be exported under the designation 'chocolate' to the other Member States. It would not necessarily allow the other Member States to make such chocolate themselves.

In such a blocked situation, we cannot but urge the EC Commission **and** the Member States, to find a solution to escape this deadlock. Either a modification of the Directive, which many are in favour of, or a relaxation of the position of the Member States, but a compromise must be found which finally allows the free circulation of chocolate in Europe. Obviously the attitude of the European Parliament will be determinant as well.

Talking of chocolate, we have imperceptibly entered the new era, 'the after-1985 period', the era of free movement of foodstuffs, where the slogan '1992' is on everyone's lips. If it is true that this so-called *new approach* has raised many questions and has caused some fears — both of these not without reason — it is also true that it represents both a considerable challenge and an extraordinary opportunity. The challenge consists of providing a coherent framework for achieving the basic objective of developing and maintaining consumer confidence in foodstuffs through the minimum necessary legislation, which fully takes into account appropriate scientific, toxicological, and safety assessments. The opportunity, for the food industry, lies in the opening-up of a community market of 337 million consumers — not to mention the creation of the European Economic Space or the newly promising situation in Eastern Europe.

On the regulatory level, according to the *CIAA*, the Confederation of the Food and Drink Industries of the EEC, the industry

> 'appreciates very much the progress made so far and will continue to provide the Commission with the support needed. It is of vital interest to the industry that the barriers to trade which still exist are removed so that the free movement of foodstuffs can be achieved. Hence the need for a well balanced and convincing horizontal EC legislation; the Industry would rather give priority to the achievement of high quality legislation than to try desperately to complete the full Harmonization Programme by the end of 1992.'

What are the principal horizontal issues on the agenda?
We have already spoken a good deal about *additives*. In the past, national regulations on additives represented, whether one liked it or not, one of the most powerful obstacles to trade. This was even more incomprehensible since all these regulations referred to public health and were more or less based on the same toxicological evaluations: the different food habits, resulting in different intake data,

do not totally explain this; the varying interpretations from one country to another of the same scientific data do the rest. Consequently, the industry very much welcomes the future EC harmonization of additive regulations, which is already well underway with the first consolidated proposal on additives. A lot of work remains to be done, but goodwill exists on all sides and work is going ahead, although sometimes too rapidly.

Industry has contributed to the work in taking stock of the current legal situation of the national levels and in identifying its needs. We hope that these needs will be taken into account when drafting the final proposals in such a way that technological progress will not be hindered.

A major difficulty with the proposal remains the legal status of the food categories used. It is now clearly understood that these categories have nothing to do with the food denominations used for labelling purposes and are intended to describe the products concerned only in the context of additives use. By taking this pragmatic approach, huge discussions on definitions of the different foodstuffs could be avoided. However, the directives have to be transferred into national law. In order not to create new distortions, it is important that the descriptions are correctly understood by all involved. They therefore should be simple and be as close as possible to the terminology used in the corresponding food sectors.

We are confident that realistic solutions can be found for the remaining problems and are looking forward to a situation where additive rules will be the same for all food operators throughout the European Community.

Homage should be paid here to the excellent partnership which exists between the EEC Scientific Committee of Food, the members of the Commission and the numerous industrial experts.

Another horizontal issue of primary importance is that of *labelling*. Important works are at present under way and though some are only starting and, consequently, are suffering from 'childhood diseases', two domains should be mentioned which preoccupy us more particularly: claims and the quantitative ingredients declaration.

We approve, in principle, the present trend towards more restrictive regulations concerning the use of *claims*. As a matter of fact, in recent years we experience increasing use and abuse of words such as 'fresh', 'natural', 'pure', 'wholesome', 'bio', 'no artificial additives', 'no artificial anything', etc. Such claims carry some long-term disadvantages because they are detrimental to other products, they create confusion in the public by generating false hopes or conceptions, they reinforce the anti-additive phobia, etc. However we are not in favour of **total** prohibition of so-called '*health claims*', i.e. claims referring to links between diet and health. We believe that certain health claims must be authorized — they can also be useful for the nutritional education of the public which we talk so much about — provided they are properly scientifically supported. We must not totally close the door on 'health claims' but, with extreme prudence, leave it ajar. This is also the conclusion that we have drawn from the experiences over recent years in the USA. In passing, we applaud the adoption of the Directive on '*Nutrition labelling*', and especially the fact that the nutrition labelling remains voluntary.

The legislator and the food industry have a common enemy: it is the power, the terrorism, of beliefs, of food myths. People believe in the superiority of 'natural foods' and preach the abandon of the modern technology of our industrialized society to return to a wild, healthy existence as in the prehistoric days.

People brandish the dangers of chemicals in foods, they are afraid of food additives. People are afraid and they frighten others — about irradiation, about biotechnology.

To lose weight, to be slim, fit, tanned and good-looking, to stay young and healthy, the advice given by doctors and scientists appears too simple. So people prefer to turn to miracle cures, towards vague ideologies, even if they don't work, even if they are dangerous, but they promise wonders!

If, in a general manner, we only see reality filtered through our own thoughts and fantasies, this is even more true where nutrition and health is concerned. In other words, we never, or rarely eat just any old thing any old how — but we eat according to our culture, to our perceptions and to our myths.

But let us get back to the question of *quantitative labelling*.

The question of quantitative labelling is something entirely different: it is proposed to make mandatory the quantitative declaration of **all** ingredients on the labels in the list of ingredients. Apart from technical difficulties like:

— the need to harmonize basic definitions across Europe (for example, what is meat?),
— the need to harmonize all quantitative methods of analysis (otherwise how is the actual quantity of an ingredient to be measured in cases of dispute, e.g. the content of fish in a fish finger),
— the absence of certain qualitative methods of analysis,

apart from these technical difficulties, the food industry is basically opposed to disclosing the full recipe of a product on the label. The path that leads a product from the fundamental research, via the development to the shelf is very long, often difficult, expensive and requires knowledge, experience and qualified people. The recipe is the know-how of the producer and this must remain so. And, finally, for the consumer's overall benefit, let us not follow those who advocate that 'more and more information equals better information'.

Finally, many questions have been asked on the function of labelling and whether it could replace compositional rules. On this subject, a single example is worth mentioning: yoghurt. Most countries in Europe and worldwide, as well as the Codex Alimentarius, have regulations restricting the denomination '*yoghurt*' to a milk product obtained by lactic acid fermentation through the action of two lactic bacteria, *Lactobacillus bulgaricus* and *Streptococcus thermophilus*. The microorganisms in the final product must be viable and abundant. This definition, which is based on ancient experience, means that yoghurt is a fresh product with a short shelf-life, with remarkable nutritional qualities which required the installation of an extremely sophisticated cold-chain for its transportation and distribution.

However, certain countries like Germany and the UK authorize the commercialization of heat-treated products under the designation, e.g. 'pasteurized yoghurt'. In our view, such a designation is misleading and should be banned: these heat-treated products have precisely lost one of the essential characteristics of yoghurt, i.e. the content of live microorganisms.

Thus, the term 'pasteurized yoghurt' or 'sterilized' abuses the positive image of yoghurt beneficial to health by, furthermore, associating it to the positive, hygienic, reassuring image of the thermic treatments. Therefore, regarding yoghurt and heat-treated products, a harmonization of EC Member State regulations is necessary: whether such an objective should be reached through a Code of Practice agreed upon by the main European manufacturers or through a Directive, remains open for the moment. One thing is certain: in this case, labelling does not resolve the problem. 'Pasteurized yoghurt'? It is like saying 'a blonde brunette'.

With the question of the 'new EC approach' to food legislation, discussions began as to whether the *CEN, the European Committee for Standardization*, should continue or even expand their activities in the food sector.

The EEC Directive 83/189, laying down a procedure for the provision of information in the field of technical standards and regulations, initially excluded agricultural products, but was modified so that these were also covered by the information procedure from 1 January 1989. Various sporadic contacts between DGIII and CEN took place in order to examine the possibilities of a contribution of CEN to the activities in the food sector. Although the outcome of these discussions is not quite clear, we understand that two possible orientations were proposed by the CEN, that is:

— the elaboration of analytical methods and
— the European standardization of finished products.

On this issue, we could subscribe to the elaboration of analytical methods — on condition that the work of the mother organization, ISO (International Standard Organization) is taken into consideration and on condition also that national analytical methods are not used as non-tariff trade barriers. However, we are not in favour of CEN standardizing finished products; this should remain the privilege of EC authorities.

Plus, if the directive has the advantage of being compulsory for everyone, the simple standard would be respected by some but not others, thus allowing unfair competition.

If references are nonetheless necessary, we propose to refer to the work of *Codex Alimentarius*. Codex Alimentarius, the Joint Food and Agricultural Organization of the United Nations (FAO) and the World Health Organization (WHO) Food Standards Programme has existed for nearly 30 years and has accomplished a pioneering work, unique and remarkable in the international harmonization of food regulations. Codex has already influenced the world's food laws and certainly will continue to do so. The toxicological evaluations on food additives, contaminants, veterinary drug residues and pesticides of the JECFA (Joint FAO/WHO Expert Committee on Food Additives) and the JMPR (Joint FAO/WHO Meeting on Pesticide Residues), have become indispensable. The creation of a forum in which members of FAO and WHO, governmental representatives, the industry, the scientific world and consumer associations meet and talk, is not the least of Codex's achievements. There is also the distinct possibility that the Codex standards may be used as a reference in a GATT (General Agreement on Tariff and Trade) Code on abolition of non-tariff trade barriers. For all these reasons, we welcome a reapproachment in the work of EC-Codex and we applaud the recent Proposal for a Council Regulation on the acceptance by the European Community of Codex standards for foodstuffs and of maximum limits for pesticide residues or maximum limits for residues of veterinary drugs in food drawn up under the Codex Alimentarius Programme.

A draft directive on *'novel food ingredients'* and *'novel food processes'* is in preparation, providing for an authorization procedure for such products.

This is a domain of great importance for the food industry in the future and the concepts upon which the text in preparation is based cause a lot of concern. In this proposal, authorization is linked with the concept of 'novelty'. Firstly, it is very difficult to define 'novelty': in the absence of clear and sound definitions, the proposal would create legal uncertainty. Secondly, 'novelty' as such is not a risk or a health factor, therefore using it as the main criterion for evaluation does not seem appropriate. Taking *biotechnology* as an example, it is clear that the use of both conventional and modern (such as recombinant DNA) techniques for the introduction of genetic modifications may give rise to hazards. Safety evaluations should therefore be made of the genetically modified organism independent of the method used to introduce the modification.

Such safety evaluations should be carried out on a case-by-case basis following a decision tree model taking into account such factors as the type of organism the sources of genetic material involved, the type of degree of control of the genetic modification, and the intended mode of exploitation.

It is understood that authorities, in order to be able to take up their basic responsibility, want to have insight into the results of the safety evaluations carried out by industry. To make this possible, it is proposed to establish a *notification procedure*. Here again it is essential that generally accepted rules exist to determine which projects should be notified and which not. Recently CIAA, the Confederation of the Food and Drinks Industries of the EEC, has proposed a scheme, based on a decision tree approach, that would enable this decision to be made. The notification procedure should be flexible and limited in time in order not to hinder the development of new products.

Even in the case of products considered for notification, we are dealing with products where no health problems are present. A notification procedure can thus not lead to a preliminary authorization step.

From the discussions at the 8th International Congress of the European Food Law Association concerning *Food law and novel foods* that was held in Luxemburg, 10–12 Oct. 1990, it was obvious that the main difficulty is to develop a legal system that is juridically sound while providing for sufficient flexibility to focus the system on the real areas where safety problems exist. It was considered that the use of the novelty criterion was not relevant.

In the absence of a generally accepted concept it is important that the problem should be further discussed and that no definite proposals should be made before the basic contradiction between legal and scientific approaches is solved.

Again the problem is not specific to the European Community. The very important conference of FAO and WHO on Food Standards, Chemicals in Food and Food Trade in Rome in March 1991 also dealt with this matter. This conference was the ideal occasion to join efforts and to develop a system that could be generally accepted.

Historically, governments dealt with the question of *food safety* by testing (i.e. collecting samples in retail outlets and having them examined), inspection and training. The food industry used a similar system, based on the examination of the end product, known as Quality Control. However, some twenty years ago it was recognized that this system was not sufficient to prevent problems, and little by little the conclusion was reached that the best means of assuring the safety of foods was the so-called 'Hazard Analysis Critical Control Point' system, the *HACCP*.

HACCP is a systematic and rational approach based on prevention, monitoring and management of food quality and safety: by identifying the critical points in the process and monitoring them with a series of defined means, a constant supervision of hygienic (and other quality criteria) is obtained.

At international level the HACCP concept has long since proved its worth: it was accepted by the WHO as an important tool in food inspection, it was incorporated by the Codex Committee on Food Hygiene in connection with its Codes of Hygienic Practices, it is recommended in the USA, and last but not least, HACCP is the basis of the modern concept of 'Quality Assurance' in the food industry.

The EC has an important hygiene programme in preparation. A general framework Directive will, it is hoped, lay down the Codex principles of application of Codes of Hygienic Practice. For specific products, mainly of animal origin, vertical directives or regulations are foreseen. Unfortunately, at present they neither include the HACCP concept nor is the relationship between the vertical and the horizontal approaches clear, that is also to say the relationship between the various EC Directorates.

Finally, it should be noted that the food industry only represents a link in the food chain which is also made up of two others: the production of the raw materials and the distribution. The authorities always had a tendancy to make the transforming industry carry the weight of regulatory constraints, but, to be efficient, these measures should be applied to the whole chain, thus creating a shared responsibility and engagement in the area of food safety. The setting up of community regulations constitutes a unique opportunity to progress in this direction.

Before concluding, some space should be devoted to a very important topic, that is, the topic of *environment*.

We consider environment to be a huge and important issue which will become increasingly important in the future. In a way, the decade of the 1990s is the decade of the protection of the environment. The food industry has always been aware of the importance of environmental issues and has taken the appropriate measures where necessary.

Together with the increase in problems there is an increase in perception: today the protection of the environment has become the attribute of all tendencies and currents and it appears on the programmes of all the political parties. Perception of environmental problems varies enormously from one region to another: it is very broad in the Scandinavian and Northern European countries, it seems to be less important in the Latin European countries and was practically non-existent — but this is changing — in the Eastern European countries, where the environmental problems are the most serious ones.

Perception also varies widely from one sector to another. In recent years attention has focused on packaging materials as a major source of the increasing municipal solid waste. However, it is essential to prevent this debate from becoming too emotional and to bring it into its right perspective. As a matter of fact, packaging accounts for 30% of household solid waste but for only about 1% of the total solid waste produced.

Packaging of food products serves a vital role in our daily lives: packaging protects products from spoilage, ensures product safety from manufacture to consumption, provides tamper evidence, communicates information, including nutritional information and serving instructions, and provides the convenience demanded by today's consumers. It is the industry's duty to develop packaged foods which correspond to consumer requirements and expectations, and which have made important contributions to the variety, convenience and safety of the food supply over the years. Therefore, we should feel free to utilize the most efficient and appropriate packaging materials available.

As mentioned before, the food industry is committed to policies aimed at protecting the environment and in particular at reducing the impact of packaging on the environment. However, we know there is no single solution to this problem. Bans and discriminatory measures like those taken, for example, against PVC (polyvinylchloride) packages in certain countries, will certainly not solve the problem which is considerably more complex and requires multiple solutions. Only a positive *integrated waste management approach* incorporating a combination of source reduction, reuse, recycling, waste-to-energy incineration and limited landfilling will minimize the solid waste problem. Successful waste management systems exist and are currently in operation in Europe and North America.

In this respect, we should like to pay credit to the creation, in 1989, of ERRA, the European Recovery and Recycling Association. ERRA comprises 23 major packaging and consumer product companies. It has established pilot projects in several European countries, thus aiming to demonstrate to legislators that efficient waste management systems are possible. Thanks to the accumulated wealth of experience and knowledge, ERRA — and together with ERRA the food industry — has become a valuable discussion and action partner.

We hope that in formulating a comprehensive European Community strategy, the Commission will build upon these existing experiences, which have also demonstrated the success of partnership between public authorities, industry (both packaging manufacturers and users, i.e. the food industry), trade and consumers. As Americans say, we want to be part of the solution and not of the problem.

What conclusions can we draw from this overview of food technology and legislation?

Progress in food technology has always been more rapid than the evolution of legislation. The discoveries are made by research and development in the food industry, under the responsibility and control of experienced men and women, who continually strive to make scientific progress. Legislation must not slow down this technological progress. Laws and regulations must be reasonable, rational, realistic, applicable and controllable. There are no doubt good and bad laws. There are, particularly, laws and regulations which reflect a certain state of preoccupation, of dominating convictions, of fashions, of mythical currents and which reflect the pressure of the media or influential groups. The authorities, the industry, can try to contend with these trends but these fears, ideologies and myths often prevail over reason.

It is true that the old days were good days — particularly when we only choose to remember the good things! But when reason dominates, let us recognize and conclude that our food has never been as rich or as healthy as it is today. Man — omnivorous par excellence — can enjoy living to eat and not just eating to live!

# I.4

# Directive 89/397/EEC:
# Assessment and outlook

**G. Verardi**
Commission of the European Communities, Brussels, Belgium

## INTRODUCTION

The 1st of January 1993 will be a milestone in European history, for on that day the internal frontiers of the European Community will have vanished. People, services, capital and goods will all be able to move freely between the twelve Member States. Products will no longer run into tax, administrative and bureaucratic obstacles. Manufacturers will reap the benefits of a market of over 300 million people. The coordination or mutual recognition of manufacturing procedures, product composition and standards will eliminate the technical incompatibilities which hamper trade. Lastly, harmonization of national laws and regulations will strenghten the guarantees offered to European consumers.

## CONSUMER PROTECTION AND THE EUROPEAN COMMUNITY

As far as food products go, consumer protection is two-pronged: on the one hand, it seeks to protect consumers' health against the adverse effects of consuming food of poor hygienic or nutritional quality; on the other, it aims to prevent the economic damage which can result from the presence on the market of foodstuffs of doubtful quality.

Until fairly recently these two features of consumer protection were the exclusive domain of the national administrations. Each administration tackled the job as it saw fit, though the usual approach in a number of Member States was basically that of monitoring:

— in the country itself, the manufacture, trading and selling of foodstuffs, and
— at the borders, the importation of foodstuffs.

Lack of harmonization had serious drawbacks. Firstly, it was impossible to ensure uniform consumer protection throughout the Community, especially since some national legislations did not even provide for monitoring of foodstuffs not intended for domestic consumption. In addition, products subject to a dispute in a Member State of destination might be held up until the problem was settled, or sent back to the consigner, with all the awkward consequences which that entails.

The removal of intra-Community barriers will put an end to any national controls liable to hamper free movement. Checks on foodstuffs carried out by importing Member States will be replaced by checks carried out by the public authorities in the manufacturing Member State on the basis of standardized Community criteria and intended primarily to ascertain whether *national* requirements in matters of wholesomeness and consumer protection are met.

## THE CASSIS DE DIJON JUDGEMENT

This a logical development of the *Cassis de Dijon* judgement of 1979, in which the Court of Justice of the European Communities ruled that any merchandise meeting the legal requirements of the manufacturing Member State is entitled to move freely within the Community, unless general mandatory requirements dictate otherwise. The judgement immediately highlighted the work of the monitoring services of the manufacturing Member States, which were primarily responsible for checking that goods complied with national legislation. One thing should be made clear: in the large single market, national monitoring authorities will not just be required to carry out checks; they will also have to try to provide the widest and most comprehensive protection possible for the consumer.

I should point out that, where necessary, Member States will retain the right to carry out checks on foodstuffs imported from other Member States. However, such checks must be carried out on the same basis as those made on domestic products intended for domestic consumption and must not occasion barriers to free trade.

All monitoring in the single market must of course be based on common principles, which must take account of the supranational dimension of monitoring at the stage of sale to the ultimate consumer, when the protection of a single citizen of a Single Member State becomes the protection of the entire Community.

## AIMS AND CONTENT OF DIRECTIVE 89/397/EEC

The Commission was well aware of all this when drawing up Directive 89/397/EEC, adopted by the Council of the European Communities on 14 June 1989, which harmonized the basic principles for the official control of foodstuffs in the Member States.

The Directive is in fact a framework, gradually to be filled in with specific provisions wherever these are felt to be necessary. It will need to be supplemented by suitable provisions, together with standards, specifications, etc., wherever these prove to be necessary.

Having said that, let me run through the aims and the content of the Directive.

First of all, the Directive covers the control of foodstuffs, additives, and materials and articles intended to come into contact with foodstuffs. It is designed to prevent risks to public health, to ensure fair commercial transactions and to protect consumers' interests, including their right to information. The Directive does not cover metrological control, since this is dealt with in other Community legislation.

Checks must be carried out both on products intended for consumption in the Member State of origin and on products intended for the single market. In principle, they must be based on the national provisions of the manufacturing Member State, except where there is (documentary) agreement with the Member State of destination. To make sure that control procedures are not evaded, Member States are not allowed to exclude a product from checks solely because it is intended for export outside the Community. (This does not mean, however, that products intended for non-member countries must necessarily be controlled in the same way as products intended for the Community market.)

Control is based on inspections, which can take place at any stage from manufacture to sale to the ultimate consumer. They may be regular or may be carried out where irregularities are suspected, though they should always be proportionate to the objective in hand.

In addition to inspections, control may involve:

(a)  sampling and analysis, the latter carried out by officially recognized laboratories;
(b)  inspection of staff hygiene;
(c)  examination of any written or documentary material.

The inspectors (who are, of course, bound by professional secrecy) are entitled, for control purposes, to take requisite measures wherever they deem this necessary. Companies have a right of appeal against these measures, if they consider them excessive, but they are nonetheless bound to assist the inspectors in the accomplishment of their tasks.

The following are subject to inspection: the state and use which is made of the site, premises, offices, plant surroundings, means of transport, machinery, equipment, etc.; raw materials, ingredients, technological aids, etc., semi-finished products; finished products; materials and articles intended to come into contact with foodstuffs; pesticides; cleaning and maintenance products and processes; processes used for the manufacture or processing of foodstuffs; preserving methods; labelling and presentation.

These principles are already widely applied in most of the Member States. That is why there do not appear to be any particular problems in transposing the Directive into national law. Transportation is well under way in most of the twelve Member States; some have even said they have no need for transpositional measures since all the principles of the Directive are already laid down in their national legislation.

## COORDINATED PROGRAMMES OF INSPECTION —
## SAMPLING AND ANALYSIS

As well as establishing these principles, the Directive lays down other operations, including coordinated programmes of inspection, starting in 1991 and for each successive year, based on recommendations from the Commission. Preliminary work has already begun.

An operation of this type and scope clearly requires suitable methods of sampling and analysis. The requirements are: specificity, precision, accuracy, repeatability, reproducibility, sensitivity (within reason, of course: for instance, there is no point in having ppb sensitivity where ppm sensitivity will do) and, finally, practicability.

The question then arises: should uniform official methods be adopted at Community level? In other words, does control within the meaning of the Directive require a Community Act to harmonize sampling and analysis methods?

The answer would seem to be yes. After all, there has already been a Regulation laying down methods of analysing wine. Moreover, the possibility already exists under Directive 85/591/EEC of 20 December 1985 concerning the introduction of Community methods of sampling and analysis for the monitoring of foodstuffs intended for human consumption.

Yet there is only a superficial similarity with that was done regarding wine. The Regulation on methods of analysing wine was adopted as part of the development of a common agricultural policy. The organization of the wine market includes special provisions of high-quality wines produced in specific areas, including the adoption of methods of analysis to check their composition and establish whether they have undergone unauthorized oenological treatment. These methods have to be obligatory for any commercial transaction; in view of the requirements of trade and the sometimes limited opportunities of traders, it was decided to allow only a limited number of standard procedures so as to produce rapid and sufficiently reliable data. So it was that methods were adopted which had already been developed by the International Vine and Wine Office (IWO) and were generally recognized.

The problem of control, within the meaning of this Directive, is rather that of removing obstacles to the free movement of foodstuffs. In this field, methods of sampling and analysis were fairly recently harmonized by Directive (until 6 October 1987: Directive laying down Community methods of sampling for chemical analysis for the monitoring of preserved milk products). However, this system has its drawbacks. In the first place, it is very costly and laborious. Secondly, it is excessively slow, to the point where a method of analysis is liable to be outmoded by the time it is adopted in a Directive. Lastly, it lacks flexibility, which petrifies the adopted methods, as it were, preventing swift adaptation to the technical progress which is so frequent in modern analytical chemistry.

There is a general consensus, then, that harmonization through Community legislation (under Directive 85/591/EEC concerning the introduction of Community methods of sampling and analysis) should be reserved for exceptional cases and that the solution to the problem should be sought elsewhere.

## EQUIVALENCE OF METHODS OF ANALYSIS IN MEMBER STATES

It would be useful to adopt a new principle whereby all methods of analysis, official or otherwise, applied in the Member States by public or private laboratories, are deemed equivalent where they produce comparable results of the same quality, on condition that this equivalence be recognized at Community level. Methods recognized as equivalent should be included in an *ad hoc* Manual, applying throughout the Community, for the protection of health and the prevention of fraud and to settle disputes of any kind.

Who would be responsible for determining equivalence?

The idea of conferring this task on one of the Commission's departments was rejected from the start, since that would have meant cumbersome administrative procedures and unacceptable costs. We have instead embarked on a course which should lead to the setting-up of a private association of food analysts. This association would work with the Commission and would be financed by it, chiefly on a contractural basis. It would include representatives from all the areas concerned (analysts from public departments and representatives of public and private research, agriculture, industry, distribution and consumers).

The Commission's role in this work would be:

— to identify problem sectors:
— to coordinate experimental work (we are currently examining the possibility of entrusting this task to the Institute for Environmental Sciences of the Joint Research Centre at Ispra);
— to supply appropriate reference materials (which could be prepared by Ispra and/or the Community Reference Bureau, acting within the Commission's Directorate-General for Research).

The association would:

— compile an inventory of all existing methods in each of the sectors identified by the Commission;
— examine them closely and, where appropriate, through experiments, on the basis of inter-laboratory tests;
— develop new methods, where needed.

Methods to be included in the above-mentioned manual would lastly have to be approved by the European Committee for Standardization (CEN).

## HARMONIZATION OF LABORATORY QUALITY STANDARDS

Clearly, though, even the most perfect of analytical methods will be of no real use unless applied by laboratories complying with suitable quality standards.

This requirement was recognized in Article 13 of the Directive on control of foodstuffs, in which the Commission undertook to report to Parliament and the Council on:

— the possibility of establishing Community quality standards for all laboratories involved in inspection and sampling;
— current provisions on the training of inspectors in the Member States;
— the possibility of establishing Community provisions on what should constitute the basic further training of inspectors;
— the possibility of establishing a Community inspection service, including opportunities for all institutions and persons involved in the inspections to exchange information.

The need to harmonize quality standards for the inspection laboratories so as to guarantee the quality of test results is obvious if we consider that:

— mutual recognition of such results is a fundamental prerequisite for barrier-free trade;
— such recognition will prevent wasteful duplication of effort;
— test results constitute the basis not only of legal proceedings in the Member States but also, under the Commission's early warning system, of appropriate Community measures.

Community provisions on principles of good laboratory practice, their application to tests on chemical products and the verification of that application have already been finalized (in Directives 87/18/EEC and 88/320/EEC). In addition, the CEN has drawn up the EN 45 000 series of standards to the same effect. All in all, this set of documents provides a valuable model for the preparation of similar provisions for foodstuffs.

With these ideas in mind, the Commission is working on a draft proposal.

## FUNCTIONS OF INSPECTION OFFICIALS

As regards the food 'inspectors' (as they are called in Article 13), their functions are not defined very clearly. A Commission survey has shown that in some Member States the term 'inspectorate' is applied to officials with a university degree, while this is not the case in others. That is why it would be better to use the more appropriate term 'inspection officer', while specifying that these officers may work at different levels and in different fields.

The same survey showed that:

— in all the Member States, guiding and organizational duties in the field of inspection are confined upon high-grade officials with university degrees, usually, but not always, in science;
— these officials rarely work in the field, the routine work being given to middle-grade officials, whose basic training, not to university level, may have been supplemented by appropriate technical training;

— management of laboratory work is always entrusted to scientists, while the actual analysis is generally carried out by middle-grade assistants.

The level of training and competance of all these people is virtually identical throughout the Community, and has proved fully satisfactory in the individual countries. In the European context, however, and with an eye to the single market, it would be useful for those in charge of inspection at the various levels to expand their knowledge by learning about:

— the structure and organization of inspection departments in Member States other than their own, and the thinking behind them;
— specific problems in each Member State and how they are being tackled;
— the rationalized techniques developed to this end;
— the principles behind the various national legislations.

This sort of knowledge cannot be gained through simple retraining courses. The Commission has instead decided to allow those in charge of inspection to gain first-hand knowledge of how things stand in the Member States by organizing and funding exchange visits between officials from the relevant national departments.

A first round has just ended. A second is being organized and should be getting under way shortly.

These rounds of visits wil be followed by seminars, which will be an opportunity for national inspection officials to meet each other from time to time and pool their experience.

A further advantage of this programme will be the setting-up of a network of personal contacts, which will be of great use in finding an easier solution to any problems which might hinder the free movement of foodstuffs in the large single market.

## A COMMUNITY INSPECTION DEPARTMENT?

Finally, there is the matter of the Community's inspection department. The Commission feels that a number of arguments support the creation of a small group of Community officials to work in the field of foodstuffs control (there are analogies in other food sectors, including fisheries, fresh meat, fruit and vegetables and wine). This group might be given an auditing role. It would contribute towards the uniform application of Community law, especially those provisions of the Treaty of Rome which deal with the free movement of goods. It could also carry out a mediating function and help solve problems arising between Member States, while at the same time facilitating the acceptance of products. It would also be important to establish certain rules setting up a system of mutual administration aid between the Member States and the Commission regarding information gained through inspections, in the way this already happens in the customs and agricultural spheres. The Commission will soon be presenting draft legislation to the Council on this matter.

## CONCLUSION

In conclusion, we must decide where specific implementing provisions need to be laid down in order to fill in the general framework of the Directive on the control of foodstuffs.

The area in which this need is felt most strongly is hygiene in the manufacture and trading of foodstuffs. Many solutions have been put forward, ranging from specific and sectoral legislation to an across-the-board approach, setting goals, establishing codes of good practice and leaving it to the Member States to choose the most appropriate means of achieving them. This question is still being studied.

# Part II
# *Nutrition*

# II.1

# Secular trends in food intake in Britain during the past 50 years

**R. G. Whitehead** and **A. A. Paul**
The Dunn Nutrition Centre, Downhams Lane, Milton Road, Cambridge CB4
1XJ, UK

A knowledge of the food intake of people in different parts of the world is crucial to our understanding of the aetiology of many diseases and to the optimization of health. The problem is that it is extremely difficult to make such measurements with any degree of accuracy and on sufficient numbers of people. Most measurement procedures are time-consuming and labour-intensive, but perhaps of even greater relevance, there is always the danger that the actual making of the measurements will place such restrictions on the subjects being studied that they will start to respond atypically and thus a false idea of the norm emerges. The more accurate one tries to be in the assessments the greater is this danger.

The paucity of meaningful dietary data for epidemiological evaluation is one of the main reasons why it is proving so difficult for Public Health Authorities, on the one hand, and the different parts of the Food Industry, on the other, to develop realistic policies towards which foods we should be encouraging people to eat. We need a better understanding of the precise link between how different dietary practices influence the way the body functions.

## THE BRITISH NATIONAL FOOD SURVEY

With all these investigative conundrums it is not surprising that investigators have had to resort to more indirect ways of getting at the information they need and in the quantities necessary. We will be describing such a data source which is used very extensively in the United Kingdom by a wide range of Government Departments as the basis for policy evaluations. This is the National Food Survey which is conducted by our Ministry of Agriculture, Fisheries and Food [1,2]. As I will be emphasizing, it is this continuity of data collection which represents one of the Survey's greatest scientific strengths.

The National Food Survey, when it started, was not intended to be a primary source of nutritional information at all. Rather it was an economic venture to ensure that all vulnerable sectors of the British population had sufficient income to purchase adequate quantities of food. It still does this, but it is now best known for the regular flow of information on the intakes of food and nutrients in Britain with time [3]. The secular trend information which emerges can be compared with similar data on disease patterns collected by the health authorities.

The sample is selected at random to be representative of the United Kingdom as a whole. Based on local authority districts, the object is to cover evenly the whole of Wales and Scotland as well as different geographical parts of England. Care is taken to ensure that rural and semirural as well as urban areas are all adequately covered. By spreading the study throughout the year, seasonal trend information is also collected. In order to determine the influence of social class and purchasing power on dietary trends the income of the head of the household is also noted.

The Survey is based on the person who is primarily responsible for buying the household food, normally the housewife but not always! This person is asked by specially trained interviewers to keep a record for seven days of the description, quantity, and cost of all food items entering the home for human consumption. The only foods which are excluded are sweets, chocolates, alcohol and soft drinks, because these items are frequently purchased by various individuals within the family without the knowledge of the housewife! This information gap can be partially filled from other government statistics on food supplies. The data are not just confined to purchased food; any free food or food grown in people's gardens is also recorded. Amongst other crucial information that is collected is a list of members of the household and visitors present at each meal, together with a brief description of the foods served.

There are of course a number of theoretical shortcomings in this type of Survey, especially from a nutritional point of view. The main one is that only food brought home is recorded; but it has been shown that this consistently represents around 85% to 90% of the total food consumed, and, furthermore, it has been determined that the basic patterns of food eaten away from home are very similar to those consumed at home. The Survey also only measures the food acquired, not that actually consumed; but from a trend analysis point of view it has been assumed that, if analysed over a sufficiently long time-frame, matters balance out. Measures of household waste as proportions of total food purchases are also made on a regular basis to ensure that no confusing variable becomes introduced from this cause. The exclusion of sweets, alcohol, soft drinks and vitamin pills also represents a problem, but with the exception of the latter this mainly affects energy-intake analysis.

Apart from producing data on the actual amounts of food and nutrients consumed within the home, the data are also compared with the dietary recommendations by the Department of Health [4] for groups of people. For this analysis, the conventional RDA is adjusted to take into account both the family structure and also those needs of some members of the household that were met from outside the home, for example in restaurants or in school or works canteens. A factor of 10% is also routinely deducted from the nutrients theoretically available for consumption to allow for waste.

We have gone into some detail to point out the shortcomings of the British National Food Survey because it is important in terms of interpreting the data that follow. May we just emphasize, however, that it is not necessarily a knowledge of the absolute amounts of nutrients that is required for epidemiological evaluation; rather, it is information on patterns of change with time which is required when one is performing regression analysis on diet–disease interrelationships. Although we would not wish to gloss over any methodological complexities, the unique value of the Survey lies in the fact that it has been operating in a disciplined and regular manner for 50 years.

### TRENDS IN ENERGY CONSUMPTION

Fig. 1 describes our estimates as to what has been happening to total energy consumption of the British population since the early 1950s, relative to the RDA as modified in the way described above. The evidence would indicate a major drop over the past 20 years or so. No one knows for certain why this has been so, but it is usually assumed to reflect a major change in lifestyles. The average British person now lives

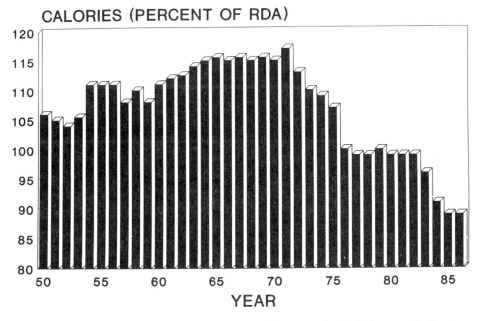

Fig. 1 — Changes in energy intake as a percentage of Recommended Daily Amount [4]. Based on the National Food Survey 1950–1986 [1,2].

in a very sedentary manner and, within limits, appetite for dietary energy does match needs. There is also a greater desire to avoid the excesses of obesity, and whilst our health surveys do not indicate any major improvement in the incidence of over-weightness perhaps the rise that might have been anticipated would accompany a sedentary lifestyle has been avoided!

## SECULAR TRENDS IN FAT CONSUMPTION

All sets of dietary guidelines for health contain advice on limiting the amount of fat consumed, especially the amount of saturated fat [5]. Fig. 2 gives the British intake data, as derived by the National Food Survey, for the latter component. Once again a definite fall is discerned over the past two decades, which could be interpreted as a positive trend *vis a vis* the advice. Unfortunately, however, if one calculates the percentage saturated fat energy, the improvement is not so impressive. By and large, people have been consuming less of the same things and saturated fat consumption has merely been falling in line with total energy.

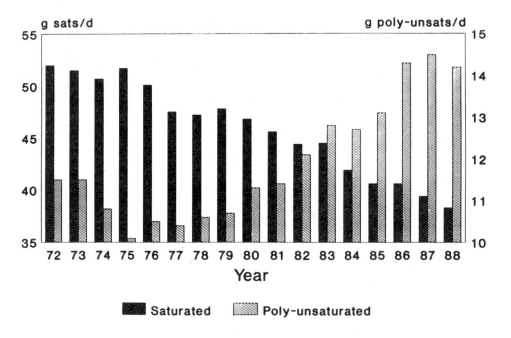

Fig. 2 — Changes in average household consumption (g/person/day) of saturated and polyunsaturated fatty acids from 1972 to 1988 [2].

Other important advice concerns a recommended increase in polyunsaturated fatty acid consumption [6]. Fig. 2 also demonstrates that after a nadir at around 1975 the polyunsaturated fatty acid content of the average British diet has been changing in line with this advice, and Fig. 3 provides the corresponding information for the polyunsaturated/saturated fatty acid ratio (P/S ratio). Although this ratio has been rising steadily, the current value reached, 0.35–0.4, is still significantly lower than it is in many countries, including the United States.

There is at present a big interest in the potential importance of the monounsaturated fatty acid component of the diet because of the relatively low incidence of cardiovascular disease in Southern Europe where the commonly used olive oil is particularly rich in this component [7]. Fig. 4 shows the consumption trends for the

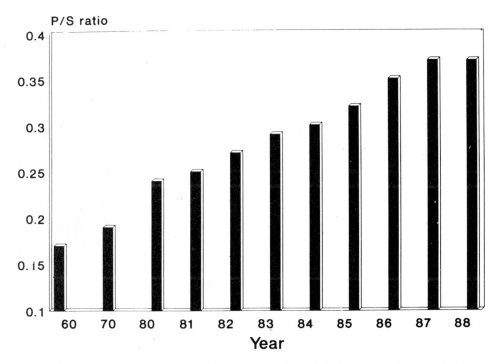

Fig. 3 — Polyunsaturated to saturated fatty acid ratio (P/S ratio) of the average household diet from 1960 to 1988 [2].

United Kingdom; but, in contrast to the polyunsaturated fatty acid component, there has been a consistent fall in the monounsaturates.

We have emphasized in this paper the importance of being able to compare these dietary trends with corresponding values for health and disease. It is especially important to be able to do this in the case of cardiovascular diseases and Fig. 5 compares changes in the mortality rates from this group of disorders for Scotland, England and the United States [8]. Some 20 years ago, mortality rates in the USA were just as bad as in Scotland, whilst England compared favourably. Since that time, however, there have been dramatic improvements in the USA, as in a number of other countries, which have not really been matched in Britain. Whether or not this absence of improvement in the UK truly reflects our dietary habits is a matter of intense discussion. Some authorities point out that the steady rise in the P/S ratio would appear to have had little or no positive effect, thus disproving the cause and effect hypothesis, whilst others would draw attention to the monounsaturated figures and the low usage of oils like olive oil in British cuisine as the confounding factor.

## FRUIT AND VEGETABLE CONSUMPTION IN THE UK

In our opening description of the British National Food Survey we described how it was designed to detect variations in consumption in different parts of the country as

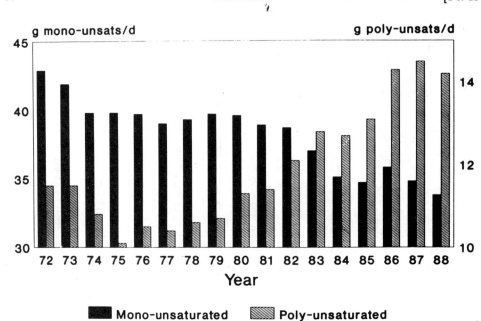

Fig. 4 — Changes in average household consumption (g/person/day) of monounsaturated and
polyunsaturated fatty acids from 1972 to 1988 [2].

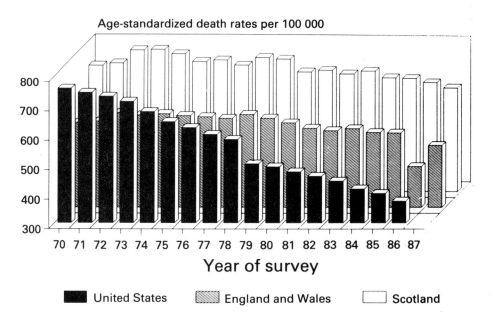

Fig. 5 — Comparison of mortality from coronary heart disease (age-standardized death rates
per 100 000) from 1970 to 1987 in the USA, England and Wales, and Scotland [8].

well as amongst different social strata. Fruit and vegetable consumption is believed to be important for a wide variety of health reasons but compliance with this advice does vary dramatically, as shown in Figs 6, 7 and 8. Compared with South-eastern England and East Anglia the consumption of fruit in Scotland is only about one half (Fig. 6) and there are similar differences in the case of fresh vegetables (Fig. 7).

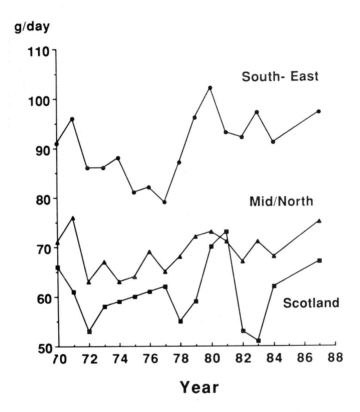

Fig. 6 — Variation in the consumption (g/person/day) of fresh fruit in three areas of Britain (the South-east, the Industrial Midlands and North, and Scotland) from 1970 to 1988 [2].

Likewise there are consistent differences relating to social class (Fig. 8), with the top 10% of wage earners again consuming around twice the amount of fruit and also vegetables. Why there should be these differences is not especially clear. It is unlikely to be due to purchasing capacity but rather to long-held dietary customs. Whatever the reason, however, it does point to thee importance of different emphases in nutritional education being required depending on geography and the social background of the people concerned. There is a tendency in the United Kingdom for nutritional education to be limited to just one type of approach. Clearly this is not adequate.

   As might be anticipated, the marked differences in fruit consumption are also accompanied by equally dramatic variations in vitamin C intake, as shown in Fig. 9.

g/day

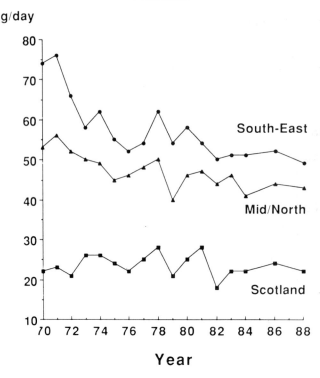

Fig. 7 — Variation in the consumption (g/person/day) of green vegetables in three areas of Britain (the South-east, the Industrial Midlands and North, and Scotland) from 1970 to 1988 [2].

A similar analysis would also show variations in other dietary components largely derived from vegetable sources, such as the β-carotenes. Understandably, with all the current interest [9] in the role of such compounds as possible antioxidants in preventing free-radical damage and thus minimizing the development of cardiovascular disease and various cancers, there has been speculation as to whether the relatively high incidence of these disorders in Scotland and amongst the less affluent might be as much linked with fruit as with fat!

The data for fruit and vegetables also demonstrate another advantage of long-term prospective measurements: it is immediately apparent that our advice about fruit and vegetable consumption is not being followed. There is absolutely no sign of those sectors of the community with low consumption statistics showing any improvement. Whilst this is a disappointing finding, at least we are left with no false ideas as to the success of our efforts!

## OTHER RELEVANT DIETARY CHANGES

The British National Food Survey is a gold-mine of data items relating to modern nutritional hypotheses and it is not possible to discuss them all here. Perhaps a further three could be singled out, however, as being of special relevance to current European interests.

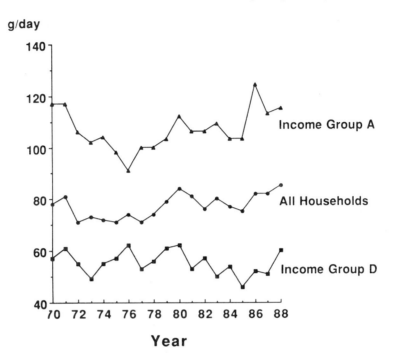

Fig. 8 — Comparison of the consumption (g/person/day) of fresh fruit in the highest income group A (more than £375 per week in 1988) and lowest income group D (less than £100 per week in 1988) with that of all households, from 1970 to 1988 [2].

Fig. 10 shows what has been happening to the British consumption of sugar within the home, the break-down being by family income. Whilst this consumption has dropped by almost one half, it is again obvious that it is the less well off who are the greatest users of this product. We will leave you to speculate on what the health significance of this might be!

Fig. 11, also, shows trends which we personally find disturbing. Milk consumption is also falling. Milk is a valuable and natural source of many micronutrients. No doubt some of these changes are due to worries by the consumer about the saturated fat content of milk, but although the National Food Survey does indicate some increase in semi-skimmed and skimmed milk product consumption this has been insufficient to affect a balance. Clearly there are grounds for more intensive and rational education in this area too.

Finally, and to end on a more positive note, there have been signs of a much improved consumption of wholemeal bread within the United Kingdom (Fig. 12), although once again it is the more affluent South-east rather than Scotland which is leading the way.

## CONCLUSION

No one could claim that the British National Food Survey represents the perfect way of defining what is happening to our diet: there are too many scientific gaps in the

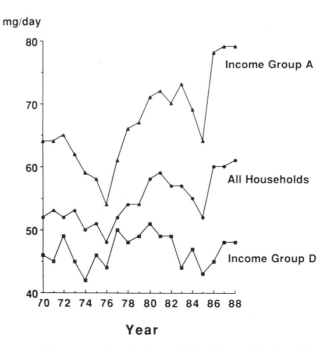

Fig. 9 — Comparison of the consumption (mg/person/day) of vitamin C in the highest income group A (more than £375 per week in 1988) and lowest income group D (less than £100 per week in 1988) with that of all households, from 1970 to 1988 [2].

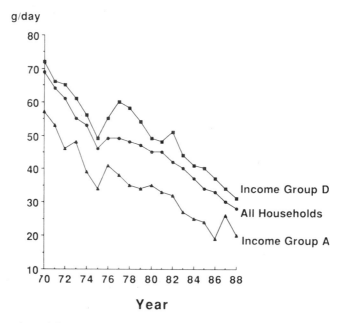

Fig. 10 — Comparison of the consumption (g/person/day) of sugar purchased for use in the home in the highest income group A (more than £375 per week in 1988) and lowest income group D (less than £100 per week in 1988) with that of all households, from 1970 to 1988 [2].

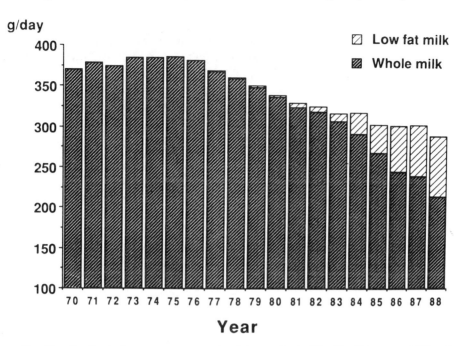

Fig. 11 — Decline in the average consumption (g/person/day) of liquid milk between 1970 and 1988 [2], showing both whole milk and low-fat milks (skimmed and semi-skimmed).

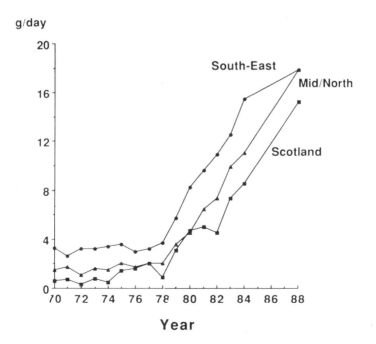

Fig. 12 — Rise in the consumption (g/person/day) of wholemeal bread in three areas of Britain (the South-east, the Industrial Midlands and North, and Scotland) from 1970 to 1988 [2].

system for that. We do hope, however, that we have been able to convince you of the scientific importance and strategic value of setting up a practical and reproducible system which enables a country to keep a close watch on what is happening to the food habits of its citizens over an extended period of time. The findings may not always be comforting, neither from a political point of view, nor for those of us who are trying to influence people's habits through educational means; but at least we know our shortcomings!

## REFERENCES

[1] Ministry of Agriculture, Fisheries and Food. Domestic Food Consumption and Expenditure 1950–1964. Annual Reports of the National Food Survey Committee. London, HMSO: 1952–1966.

[2] Ministry of Agriculture, Fisheries and Food. Household Food Consumption and Expenditure 1965–1988. Annual Reports of the National Food Survey Committee. London, HMSO: 1967–1989.

[3] Derry, B. J. and Buss, D. H. The British National Food Survey as a major epidemiological resource. *Brit. Med. J.* 1984: **288** 765–7.

[4] Department of Health and Social Security. Recommended daily amounts of food energy and nutrients for groups of people in the United Kingdom. Rep. Health & Soc. Subj. London, HMSO 1979: No. 15.

[5] Department of Health and Social Security. Diet and cardiovascular disease. Rep. Health & Soc. Subj. London, HMSO 1984: No. 28.

[6] National Advisory Committee on Nutrition Education. Proposals for nutritional guidelines for health education in Britain. London, Health Education Council 1983.

[7] Grundy, S. M. Monounsaturated fatty acids and cholesterol metabolism: implications for dietary recommendations. *J. Nutr.* 1989: **119** 529–533.

[8] World Health Statistics Annuals 1970–1987.

[9] Willett, W. C. Vitamin A and lung cancer. *Nutr. Rev.* 1990: **48** 201–211.

# II.2

# The consequences of 100 years' evolution of dietary habits in Europe with regard to nutrition

**Rudolph J. J. Hermus**
TNO Toxicology and Nutrition Institute, PO Box 360, 3700 AJ Zeist, The Netherlands

## INTRODUCTION

One hundred years of development of industrial food production and consumption may seem a long period. It covers about three generations. However, before this period of recent change, mankind has gone through almost 60 million years of evolutionary phases. From the hunting and gathering by the first monkeys (60 million years ago) and by *Homo erectus* (about 2 million years ago), the switch to the first agriculture by *Homo sapiens* (about 8000 years ago) is rather recent [1]. On this scale, 100 years represents only a few counts. However, probably never before in history has food procurement so dramatically and so drastically changed as during the last 100 years. As a result of important discoveries in the areas of biology and science, the science of human nutrition developed, as well as food science and technology. Progress in these scientific domains has had enormous impact on food provision and food quality as well as on public health and disease patterns and ultimately even life expectancy.

Four eras of nutrition research can be distinguished [2]. First came the era of discoveries of the existence of nutrients, up till about 1940. Secondly, during the forties, there was a transition period, during which nutritional knowledge was successfully applied in food rationing. At the end of this period it was believed that virtually everything about nutrition which was worthwhile was now known; recommended dietary allowances (RDAs) are the fruits of this period. However, in the fifties and up till the present time, a third era developed in which biochemical functions of nutrients were investigated in depth, which yielded evidence of a role for nutrients in numerous bodily functions. This led to the transition to the fourth era, an era of preventive nutrition, during which relationships are demonstrated between dietary habits and health and disease patterns. The application of epidemiological methods to the study of these relationships has demonstrated the enormous potential of this area. However at the same time, because of the inconclusive nature of many epidemiological studies, a lot of controversies have also arisen. Dietary guidelines for the population are the condensed form summarizing the collective knowledge. It can be anticipated that in the near future, as soon as molecular biology permits us to characterize the biochemical individuality of people, a fifth era will emerge called individualization of nutrition. Application of information technology will make it possible to match, more or less perfectly, individual needs and requirements with properties of foods and drinks, including the sensory properties.

## HISTORY

The science of nutrition has always been linked with progress in the sciences of medicine, physics, chemistry and physiology [3]: Lavoisier, Liebig, Mayer, Joule, Atwater, Boussingault, Mulder, Bernard, Pasteur are only a few examples. It was in 1890 that Eijkman in the Dutch East Indies, in search of a 'germ' which was thought responsible for the serious and rampant beri-beri disease, put forward the idea that neither a germ nor a toxic factor probably was to be held responsible but a 'partial hunger', a shortage of something. It completely upset the scientific community that the absence of a factor might cause disease. Mortality in history has been studied carefully by Cairns [4], among others. He compared age distributions at death from inscriptions on gravestones of paleolithic man (40 000 BC) and neolithic man (10 000 BC) from Morocco and Hungary with life tables from contemporary hunter-gatherers in the Kalahari desert, the Kung tribe. It appeared that the life curves were essentially similar. At the age of five years, life expectancy was about 40–50 years. Maximum life expectancy was about 80 years. In ancient Rome, life expectancy was much worse, probably owing to adverse hygienic conditions and a high population pressure. Survival was close to survival of objects with a half-life of 14 years. Survival in England and Holland in about 1850 was not essentially different from that of early agriculturists. Life expectancy at birth in the Netherlands was 36 years in 1840, 51 in 1900, 71 in 1950, 73 in 1985. About two-thirds of mortality was caused by or related to some microorganism. Airborne microorganisms were the most important single cause of death, followed by water and foodborne microorganisms. Diseases associated with affluence, such as gallstones, renal stones, and cancer of the breast, ovary and prostate increased. Diseases associated with poor living standards decreased. These include stroke, stomach cancer and rheumatic heart disease. They are still more prevalent in the least affluent areas and in people with the lowest incomes.

A third group of diseases is at different times associated with both affluence and poverty. It includes ischaemic heart disease, obesity, appendicitis and duodenal ulcer. In the first part of this century incidence of these diseases rose and they were more prevalent among the rich. Subsequently, incidence fell and they became more common among poorer people.

The importance of nutrition as a determinant of health and disease is illustrated by the share of the total health care costs which is taken up by nutrition-related diseases and disturbances. In the F.R. Germany it was calculated, on the basis of figures from 1980, that from total direct health care costs of $87 \times 10^9$ DM about 33% or $29 \times 10^9$ DM were related to nutrition-associated morbidity [5]. Among those diseases tooth decay was the most expensive, followed by ischaemic heart disease, hypertension, diabetes and cerebrovascular disease.

## DIETARY HABITS IN WESTERN EUROPE: 1850–1990

The dietary habits in Europe reflect the different socio-economic and technological phases of its history. Between 1800 and 1870 the first industrial revolution took place. Potatoes and rye were the main staple foods. About 65% of the household budgets went into food. From 1870 till 1920 the second industrial revolution occurred. Wheat became the staple grain. More food energy became available, but still about 60% of the budget was spent on food.

Between 1920 and 1950 society was restructured. Meat and also sugar became more widely available. About 50% of the budget was used for eating and drinking. From 1950 till the present time the economy assumed the form of mass production. Incomes and wealth rose sharply. More food energy became available, especially from products of animal origin. The proportion of the household budget spent on foods dropped from 30% even to below 20% in certain countries. Food became relatively inexpensive. This last point is also illustrated by the trends in the working time required to earn 1 kg of specified foods. In 1885 it took 120 minutes to earn 1 kg wheatbread; in 1920 it took only 60 minutes, in 1940 30 minutes; and presently it takes just 10 minutes. To earn 1 kg potatoes it took 15 minutes in 1885 but takes only 1 minute at present. In order to earn 1 kg beef meat 360 minutes were required in 1885 and 240 minutes in 1920; even today 120 minutes are required, demonstrating that meat is still relatively expensive.

Interesting data about dietary trends in several European countries have been described recently [6,7,8,9].

Recent trends in The Netherlands and Belgium have been described as well [10,11].

A very comprehensive description of the gross consumption of foods per head per year has been published for the F.R. Germany [6,7]. It covers the entire time span since the beginning of industrialization (1850–1975). A more detailed description of recent trends is provided by the *Ernährungsbericht 1988* [5]. Table 1 shows an overview of the most significant changes in the German diet over time. Among the losers are: rye flour, potatoes, pulses, and generally plant products. Among the winners are: fruits, citrus fruits, sugar, oils and fats and total meat, especially of pork and poultry origin. The yearly fluctuations accurately reflect the economic and political periods of instability. The milk group appears rather stable. However, within this group a considerable decrease has occurred in the drinking of fresh milk and an increase has been demonstrated for cheese consumption from about 5 kg in 1950 to 15 kg in 1985.

A description of trends in the diet of France has been published by Dupin *et al.* [8]. Similar trends as in Germany were observed (Table 2). Bread, potatoes and

**Table 1** — Gross food consumption in Germany between 1850 and 1985
(kg/head/year) [5,6,7]

|              | 1850 | 1890 | 1930 | 1950 | 1975 | 1985 [5] |
|--------------|------|------|------|------|------|----------|
| Rye flour    | 56   | 66   | 57   | 35   | 14   | 13       |
| Wheat flour  | 26   | 52   | 48   | 62   | 47   | 52       |
| Potatoes     | 138  | 228  | 171  | 186  | 92   | 62       |
| Pulses       | 21   | 7    | 2    | 2    | 1    | 1        |
| Vegetables   | 37   | 58   | 52   | 50   | 67   | 66       |
| Fruits       | 15   | 23   | 30   | 43   | 81   | 97       |
| Sugar        | 2    | 8    | 23   | 28   | 36   | 37       |
| Oil+margarine| 5    | 2    | 10   | 10   | 13   | 15       |
| Beef         | 7    | 13   | 14   | 11   | 21   | 18       |
| Pork         | 8    | 17   | 27   | 19   | 45   | 51       |
| Poultry      | 2    | 1    | 2    | 1    | 9    | 10       |
| Total meat   | 22   | 38   | 49   | 36   | 83   | 81       |
| Milk products| 268  | 314  | 386  | 291  | 345  |          |
| Plant products| 309 | 455  | 408  | 429  | 379  |          |
| Animal products| 298 | 369 | 460  | 353  | 459  |          |

pulses are again the big losers, while sugar and meats are the big winners. The preponderance of foods from animal origin in the French diet is illustrated with the proportion of protein which is derived from animal products. This has increased from 27% one hundred years ago to about 70% at present.

**Table 2** — Gross food consumption in France between 1850 and 1980 (kg/head/year) [8]

|                              | 1850 | 1880 | 1925 | 1935 | 1955 | 1975 | 1980 |
|------------------------------|------|------|------|------|------|------|------|
| Bread                        |      | 219  |      | 118  |      | 68   | 63   |
| Potatoes                     |      |      | 178  |      | 125  | 91   | 85   |
| Pulses                       |      |      | 7    |      | 3    | 2    | 1    |
| Sugar                        | 3    |      | 19   | 22   | 26   | 36   | 36   |
| Meats                        | 20   |      | 41   | 47   | 60   | 90   | 110  |
| Protein from animal origin (%)|     | 27   | 39   | 45   | 51   | 64   | 71   |

For The Netherlands, Anneke van Otterloo has described in a historical–sociological study the changes in eating habits from 1840 to 1990 [9]. The most striking change which has taken place has been the development from scarcity to plenty. Production and distribution have made available a stable supply of food products. However, people have become alienated from the agricultural system and have become dependent on a complex network of human interactions as well for their food supply. Food procurement has been rationalized and made the subject of scientific endeavour. Food control is entrusted to specialists and food education has become a Government task. Food preparation, too, has changed intensely. Energy sources for cooking are readily available as is a wide variety of utensils and cookery books. The choice of basic materials or preprocessed foods has been enormously enlarged. In 1840, people had no choice. In 1990, there is such a vast range of products that people can hardly make a choice without information and education, i.e. food and ingredient labelling. Mistrust concerning adulteration has been transformed into mistrust concerning additives and environmental contamination.

The trends in food consumption in The Netherlands and also in Belgium have been summarized recently [9,10]. They confirm broadly the trends in Germany and France. Consumption of potatoes has stabilized since 1965. However, use of processed potatoes has increased from 7% in 1965 to 30% in 1987. Vegetable use has increased since 1965 by 45%, both the traditional and the novel types. Fruits showed an increase of 26% since 1965, this mainly being an increase of imported, (sub)tropical fruits like kiwifruit, melon, peach, grapefruit, orange. Within the dairy group a drastic shift took place, from full-cream milk to half-cream or skim milk and buttermilk. Full-cream milk consumption was reduced by 60% in the period 1965–1987. Cheese consumption increased by about 70% over the last 20 years. Despite all the efforts of the consumer to reduce dairy fat intake, overall consumption of dairy fat remained unchanged. In The Netherlands also, meat consumption (deboned) has increased tremendously over the years. In 1950, 26 kg meat per person was consumed (12 kg beef and 14 kg pork). In 1975, this had increased to 59 kg (20 kg beef, 32 kg pork, 7 kg poultry). The dominating position of pork has been maintained as has the increase in poultry consumption.

In Table 3 the contribution of different food groups to fat consumption has been summarized for different age groups in The Netherlands. It is evident that the animal meat and dairy products which originally were important because of their high nutrient density now also have become important sources of dietary fat. Together they account for about 40% of the fat intake.

## DIET AND HEALTH

Populations in Western Europe have never been so healthy and have a longer life expectancy at birth than ever before [11]. As a consequence, a larger proportion of the population will reach a higher age. In 1900 only 4% of the population was over 65 years. In 1985 this was 12%. In 2010 a figure of 15% is predicted and in 2035 of 24%. It is no longer considered worthwhile merely to extend life by years. Life extension should also take into account the quality of the added years involved. Therefore —

**Table 3** — Contribution to fat consumption by age groups from several food groups
in The Netherlands (%) [9]

|                          | Age (years) | | | |
|--------------------------|-------|-------|-------|-------|
|                          | 1–9   | 10–21 | 22–49 | 50–74 |
| Visible fats             | 28–30 | 30–34 | 29–32 | 33–38 |
| Meat and meat products   | 17–18 | 18–20 | 20–22 | 19–23 |
| Milk and milk products   | 13–20 | 9–10  | 9     | 8–10  |
| Cheese                   | 4–5   | 5–7   | 9–10  | 8–10  |
| Nuts/seeds/snacks        | 5–7   | 7–9   | 8     | 4–5   |
| Pastry/cakes             | 6–7   | 6–8   | 5–8   | 7–10  |

and to reduce health care costs — health policy aims at postponement of morbidity, leading to compression of morbidity in the final phase of life. This would mean that the survival curve of a population, with the quality of life taken into account, should be as rectangular as possible. Prevention strategies have been formulated with an important role for nutrition as a preventative tool to combat certain life-threatening or quality-of-life-diminishing diseases. Experts, however, have a different opinion from consumers as to what actually are important diet-related health risks. Whereas experts consider nutritional behaviour the primary factor and bacterial contamination and natural toxicants as the next two most important factors, consumers perceive environmental contamination as the primary factor and additives and behaviour as the next. Nutrition education and information here have a task to balance the views better. Otherwise people will follow the wrong tracks and fall into food faddism and quackery, with their exciting promises.

Both in The Netherlands [9,10] and in the USA and also by WHO, dietary guidelines have been published in order to formulate a dietary pattern which is most conducive to good health. All these guidelines have a remarkable conformity: they do not stress the importance of the traditional nutrients but of a dietary pattern with less saturated fats and less cholesterol and commensurate with normal body weight. The traditional Mediterranean diet would be the optimal diet at present. Even elderly people would still benefit from a change to such a diet [11]. In The Netherlands it is estimated that about 20% of the people eat in conformity with the fat guideline. If all seven dietary guidelines are considered together, less than 1% of the population conforms to this. So there is ample room for improvement, both at the food supply side, increasing real variation and providing information about the contents of a food through nutrient labelling, and at the food selection side, by the consumer's making a more informed and a healthier choice.

However, if we consider eating in the hierarchy of psychological needs according to Maslow, we should realise that eating has moved from the fulfilment of basic needs such as hunger, thirst and appetite, to the top of the need hierarchy, which is characterized by esteem, respect and self-actualization. Health promotion has a much lower value, which is probably bypassed by many.

## CONCLUDING REMARKS

Nutrition research in the near future will mainly act in the areas of overlap between dietary habits, ageing and chronic diseases. Molecular biology will help to proceed from population-based knowledge to individual susceptibility. Public health recommendations will become more and more individualized.

Food supply will have to respond to this development by greater diversity in food composition and portion sizes and by more detailed labelling of ingredients and nutrients. Traditional classification of foods will become obsolete. Foods will become more functionally designed and tailor-made, also as a result of novel technologies. Considerable research efforts are needed to elucidate: the molecular and cellular function of nutrients; the mechanisms underlying the relationships between diet and disease; the socio-behavioural aspects of dietary change.

Nutrition as a science is no longer novel. It has a long tradition; and it has matured considerably over about the past 200 years.

## REFERENCES

[1] Haenel, H. (1990) Ernährungsverhalten im Wandel der Zeiten. *Ernähr. Umsch.* **37** 235–239.
[2] Erdman, J. W. (1989) Nutrition: Past, Present and Future. *Food Technology* Sept. 220–227.
[3] Hermus, R. J. J. (1987) De Geschiedenis van de Voedingsleer. *Voeding in de praktijk* **IIA** 1–18. Bohn, Scheltema, Holkema, Utrecht.
[4] Cairns, J. (1989) The History of Mortality. In: K. K. Carroll (ed.) *Diet, Nutrition and Health*. McGill-Queen's University Press, Montreal, Kingston, London, Buffalo, pp. 309–344.
[5] *Ernährungsbericht* (1988) Deutsche Gesellschaft für Ernährung e.V., Frankfurt am Main.
[6] Teuteberg, H. J. and Wiegelmann, G. (1986) *Unsere tägliche Kost*, 2 Auflage. F. Coppenrath Verlag, Münster.
[7] Teuteberg, H. J. (1979) Der Verzehr von Nahrungsmitteln in Deutschland pro Kopf und Jahr seit Beginn der Industrialisierung (1850–1975). *Arch. Sozial Gesch.* **9** 331–388.
[8] Dupin, H., Hercberg, S. and Lagrange, V. (1984) Evolution of the French Diet: Nutritional Aspects. *Wld. Rev. Nutr. Diet.* **44** 57–84.
[9] *Voedingsbericht* (1990) Voedingsraad. SDU Uitgeverij, Den Haag.
[10] Hermus, R. J. J. and Claesen, H. (1987) De Voeding in Nederland en België: nu en in de toekomst. *Voeding in de praktijk* **IIB** 1–22. Bohn, Scheltema, Holkema, Utrecht.
[11] Kok, F. J. (1990) Voeding en Vergrijzing: Lang zullen ze leven. In W. v. Dokkum and D. G. van der Heij (red.) *Voedsel in Beweging*, Pudoc, Wageningen. pp. 224–231.

# II.3

## Developing trends in food regulation in Europe

**Jean Rey**
Hôpital des Enfants Malades, 149, Rue de Sèvres, 75743 Paris Cedex 15, France

The Treaty of Rome provided that there should be freedom of movement for people and goods in the European Economic Community. The Single European Act prescribed when it should be achieved by. The internal market is scheduled for completion by 31 December 1992. Each day brings us one step nearer.

The basic provision is Article 30 of the Treaty, which prohibits all quantitative restrictions on imports and exports, and all measures having equivalent effect, which means any commercial rules in a Member State which are likely to hinder, directly or indirectly, now or in the future, intra-Community trade. The only exceptions allowed are those prescribed by Article 36 of the Treaty 'justified on grounds of ... the protection of health and life of humans, animals or plants ...'.

The judgement delivered by the Court of Justice of the Communities in the *Cassis de Dijon* case (20 February 1979), confirmed by the *Gilli* judgement of 26 June 1980, provided the Commission with an opportunity to make its position clear in a Communication published in the Official Journal of the EC of 3 October 1980. Concerned at the increasing number of measures which were actually restricting the free movement of goods, the Commission based itself on the very general definition of obstacles to free trade propounded by the Court to conclude that any product lawfully manufactured and put on the market in one Member State must, in principle, be admitted onto the market of any other Member State.

The Commission's position was subsequently reinforced by a series of Court judgements, notably those regarding the designation and definition of products, additives, labelling, advertising, outer packaging, preparation and wrapping, designations of origin, and certification control. Hence, the case law of the Court of Justice, the only body empowered to interpret the Treaty and instruments enacted under it, and hence to declare law, has become an integral part of the corpus of Community law. Rules must apply to domestic and imported goods without distinction. They must be necessary to comply with mandatory provisions, assessed on a case-by-case basis. The rules must be in proportion to the results to be achieved and the burden lies on the Member State to show that it has no less restrictive way of achieving the same result, notably through appropriate labelling.

The judgement in the so-called 'pure beer law' case is of particular importance in this regard. The scope of the judgement goes far beyond the established precedent, as the Commission clearly emphasizes in its Communication of 24 October 1989. Henceforth, the Commission's strategy is to combine the adoption of harmonized rules with the principle of mutual recognition of national rules and standards. The Commission will be proposing harmonized rules for the food industry only in areas relating to protection of public health, consumer protection, fair trading and protection of the environment. As a general rule, these will be horizontal provisions only, applicable to foodstuffs in general and particularly referable to additives, pesticide residues, materials and objects in contact with food, labelling, presentation and preparation of products. Where no harmonized rules exist, Member States have authority to lay down rules for their own products. On the other hand, they must allow foodstuffs lawfully produced and marketed in the other Member States onto their markets. In conclusion, the Commission noted that articles 30 and 36 of the Treaty were directly applicable in Member States' national legal systems, with the consequence that any court or tribunal sitting in proceedings within its jurisdiction had a duty to apply the provisions in full as interpreted by the Court of Justice, and should not apply any national provision in conflict with them, whether pre- or post-dating the Treaty.

The principles have been laid down. They are very sound and it would be presumptuous of me to throw them into question here. But the real difficulty resides in applying them, striking the happy medium between matters which are the responsibility of the Council and Commission, and what must remain within the powers of Member States, subject to the Treaty and case law.

At a time when the European authorities — and the Commission in particular — are readily criticized on all sides for undue interference in the regulation of intra-Community trade, thereby seeking to shift the responsibility for particular difficulties onto other shoulders, I feel it worth reiterating that freedom of movement for foodstuffs is the product of an expression of political will by the Member States in signing the Treaty and Single Act. The Commission is therefore merely the instrument of that political will and we should not complain about its playing to the full the role assigned to it.

But the political will for European integration is not only that of our predecessors. It must be sustained day after day in the discussions on Directives, where so required; the demands of certain national delegations aiming, in a certain number of instances, no more than to strew obstacles in the path of freedom of movement for goods, thereby delaying its achievement. On the other hand, it is essential the letter and spirit of the Treaty be adhered to without seeking, as can be seen in the case of the draft baby milk Directive, to enact provisions whose effect will be to regulate exports to third world countries. Whatever the pressures brought to bear by lobbies to the Parliament and Commission, the principles of the International Code of Marketing of Breast Milk Substitutes approved by all Member States as a Recommendation should be strictly adhered to, by which I mean applied unreservedly but without overstepping the interpretation placed on it by the World Health Organization itself.

Finally, before coming to the substance of the matter, I feel it essential to emphasize at the outset that the necessary precondition of an integrated Europe is a questioning of the organization of national consultative authorities. We cannot press forward with an integrated Europe equipped with its policy instruments while at the same time maintaining the existing systems in each Member State without redefining the tasks of those systems and essential cooperation at a higher level.

Freedom of movement of foodstuffs within a single internal market is inconceivable unless the consumer is properly informed. Therein lies the importance of labelling, which provides him with all the information he needs without at the same time overwhelming him in a morass of figures which, rare exceptions aside, signify nothing to him. In this regard, the expression of energy value in joules is a luxury we could well do without. Indeed, why apply the International System to energy, but not to calcium, sodium and virtually all the other information printed on labels? One can only be thankful for the wisdom shown by the Directive on Nutrition Labelling in linking the obligation of nutrition labelling to the making of nutrition claims by manufacturers. Let me stress, however, that the reference nutritional intakes contained in the annex to the Directive for the application of the provisions relative to 'recommended daily allowances' are included merely as precautionary measures and could well be changed before long. The Commission has, in fact, referred this matter to the Scientific Committee for Food which is in the midst of discussions not only on reference values but on the very concept of recommended intakes or allowances which, as we know, are for use by national governments and mass caterers and meaningless when applied to individuals.

It might be useful at this juncture to refer to a number of concepts without which no coherent nutritional policy can be conducted. The first is the adjustment to virtually everything we consume, which is why we retain, very broadly speaking, just the amount of calcium, iron or ascorbic acid we need. Even a series of disorders may be regarded as the consequence of a sort of adjustment to a constitutional maladjustment, such as familial hypocholesterolaemia caused by lack of low-density lipoprotein (LDL) receptors or certain cases of obesity caused by lack of diet induced thermogenesis.

Variability and heterogeneity are two other universal concepts whose nutritional applications are unvarying. Variability conditions our entire thinking on recommended intakes, and heterogeneity allows us to understand why certain individuals are 'sensitive' or 'hyperresponder' while others are 'resistant' or 'hyporesponder' to particular constituents of food, salt and cholesterol in particular. The results of recommendations made to reduce the risk of high blood pressure resulting from excessive consumption of salt or hypercholesterolaemia related to excessive consumption of saturated fats depend on the percentages of each type of individual, the probability of those recommendations proving effective rising with the number of sensitive individuals in a population. That should encourage us to look for guideposts to particular sensitivity to specific nutrients and adopt an individualized preventive policy which is not, despite all appearances, incompatible with a mass prevention policy.

If the Council Directive of 24 September 1990 on nutrition labelling for foodstuffs should not, at first sight, pose any major problems as regards implementation, the same cannot be said of the draft Directive on nutrition claims made on labels, the presentation of foodstuffs and advertising in connection with them. France's experiences in the campaign against the harmful effects of tobacco in fact shows that advertisers have circumvented the legislation time after time, and that there has never been so much advertising for tobacco and alcohol as there is today. I am not denying that health warnings to consumers are carried in legible characters on labels, posters and magazine advertisements. No-one, therefore, can be unaware that alcoholic beverages must be 'taken in moderation' and that tobacco may damage their health. Nor is there any direct advertising for cigarettes, but cigarette brand names can be seen everywhere on advertising for lighters, clothing and rallies, not counting the financial support given by Marlboro, Camel, Gitane and the rest to Formula 1 drivers and teams. Clearly, advertisers are more inventive than the legislature.

We should also realize that in the field of nutrition and food, which concerns us here, it is more invigorating to construct positive messages encouraging people to eat or drink a particular product rather than gloomily counselling them not to. The butter promotional campaigns ('Butter or just plain?', for example) are particularly attractive in this regard, the epitome of excellence being the 'Butter adds spice to life' campaign seen throughout Paris, including in the underground, a few years ago. And whatever opinion one may have of the benefits of drinking bottled mineral waters, it surely gives pause for thought when, on opening any women's magazine one finds their weight-reducing virtues being propounded by indescribably beautiful young girls diving into pools of E.. or C... water, or quaffing C.. 'light' straight from the bottle, their long, slim legs barely concealed by a miniskirt?

It would be self-deluding, therefore, to think that a Directive could cover all probable situations and regulate matters of this order on a long-term basis. The very underlying philsophy of the issue involved is not clear. What, for example, is the point of a claim that something has a low or reduced sodium content, if that claim is also applicable to products naturally low in salt provided their sodium content has been reduced by 30 or 60%? Can we define with any precision a 'reference food', the composition of which changes with fashion, in accordance with national and regional preferences, in line with technological developments and advances in genetics? Could there be, for instance, such a thing as a reference Irish stew? While the rules contained in the preliminary draft Directive concerned may be perfectly suited to certain categories of foods, a number of which, moreover, should be dealt with in specific Directives, it would be rash at the present time to endeavour to make them generally applicable. What is required, therefore, is not a Directive but simply broad guidelines usable by Member States as a basis for an improved definition of rules in the matter, open-ended enough to adapt to the inevitable changes in the market.

We should also be turning our minds in the coming months to foods intended for special medical purposes. There are a number of problems in this regard. The first is defining what they are — a matter on which consensus has not yet been reached, albeit the definition contained in the preliminary draft Codex standard is an acceptable compromise. The tendency in America, in fact, is to consider such foods as being defined by their being medically prescribed regardless of their composition and intended use, whereas the French position is more restrictive: it is their special composition and possibly their intended use which makes them available on prescription only.

The substance of the problem, however, which has never to date been directly confronted, is that of the import of the markings on labels and warnings about the possible hazards of these 'medicinal foods' being consumed by individuals for whom they are not intended. The draft Codex standard confines itself to a statement of general principles so vague that they could equally well apply to any category of products intended for particular nutritional uses.

If the thing is to be done properly, there is in fact a choice of two positions. The first — a maximum approach — would be to say that these products, which are 'to be taken only under medical supervision', are obtainable only from a pharmacy or dispensing chemist and only by prescription. Such would be the case, for example, with products intended for the treatment of phenylketonuria, and that is what was recently recommended by the Scientific Committee for Food with regard to very low calorie diets. The second, which to me seems the minimum requirement, would be to prohibit all public advertising of this type of product. My suggestion would be: (1) to call for a prohibition on public advertising, with no exceptions; (2) to demand that these products should be obtainable in pharmacies on prescription only where a health risk would be incurred by their being consumed by individuals not suffering from the illnesses, disorders or pathologies for which the product is intended or where they may be harmful to reproductive functions, foetal development or breast feeding.

The status of enriched products and nutritional supplements may raise problems of even greater complexity than those of 'medicinal foods'. The fact of adding to a food vitamins which should normally be present in it, of 'restoring' as it were their nutritional value, should not by itself be enough to categorize it as a dietetic food, and only the fact of the addition should be permitted on labelling unaccompanied by any claims in respect of it. To do otherwise would be to encourage manufacturers not to pay the utmost attention to the quality of their raw materials and consumers to choose such products out of preference although no real benefit for their health could be expected from them. It may be, on the other hand, that certain unbalanced diets, illnesses or even just certain critical periods of life (pregnancy, breast feeding, rapid growth spurts, the elderly) fully warrant the consumption of foods genuinely enriched with particular minerals (iron, especially) or vitamins. The Scientific Committee for Food is to put forward proposals on this matter in the coming months and, without wishing to preconclude as to its opinions, it seems to me that the supplements in question must, in such cases, be of significant levels, as was recently proposed for nutritional supplements intended to compensate the protein, mineral and vitamin deficiences incurred by low calorie diets. If the consumer is not to be abused, however, would it not be advisable to define such high nutritional value foods as much by their intended use as by their composition? That is a question which must be answered.

Different again, finally, is the problem of nutritional supplements not presented as foods but in a form (tablets, pills, capsules, etc.) more closely akin to medicines. In

France at least, what determines a medicine is as much its presentation as its reputed properties and the claims made on its label. But the official authorities are reluctant to come to a decision, the government departments which could accept responsibility passing the matter on from one to another, and the courts deciding now one way, now another in litigation between pharmacists and supermarkets. For the problem is first and foremost a commercial one — the virtues of these nutritional supplements being blazoned in uncontrolled public advertising, more concerned with the miracle-working, 'wonder product' aspect that established scientific truth. But no headway can be made in this particular debate without a definition of food; a need more acutely felt, furthermore, in that additives are defined in relation to foods!

The crux of the matter on the final lap to the single market, however, lies not in labyrinthine discussions on purely grammatical concerns. No Directive will ever stop children eating sweets, neurotics from stuffing themselves full of pills and sedatives — and that is how it should be. Our efforts must be directed more towards simplification of procedures, the strengthening of voluntary codes of practice for the industry, and the introduction of a suitable supervisory system. The national authorities are still running around in circles corroborating or calling into question the opinions of the JECFA or the Scientific Committee for Food and, basing themselves on acceptable daily intakes (ADI), issuing case-specific authorizations for the use of some additive or other in some category of food or other. But it is a futile occupation, even were there no single European market approaching.

Safety in this area depends on having at least an approximate idea of daily consumption levels, which is the only way of determining whether an ADI is likely to be, or has been, exceeded and whether steps need to be taken or not. For that reason, France has recently set up an observatory on consumption as a prelude to identifying the real problems. But dietary habits and culinary traditions differ widely between Member States, and the free movement, in a single market, of goods 'generally accepted as harmless' from this viewpoint (to borrow an expression used by the FDA to describe a number of substances) requires that similar observatories should be established in all Member States, should work in concert, and forward the necessary information to the Commission. A new organizational arrangement needs to be invented — and an ethic accepted. Let manufacturers play their role in innovation, but let them also put forward cogently argued files with specific data sheets on all new substances. And let the experts, for their part, examine them promptly but with an open mind and asking only for such additional information as is strictly necessary to the preparing of their opinions. That mutual respect is the precondition of European integration.

# II.4

## Summary report — Nutrition

**B. Mathioudakis**,
DGIII, EC Commission

This part II consists of three chapters which have developed the themes of the evolution of food habits within the last 100 years and the consequence of this evolution as regards nutrition and have given some prospective views regarding food regulation in Europe.

A number of main points stemming from these chapters were identified and discussed in a round table.

### FOOD SURVEYS

National food surveys have been used for some time to monitor food consumption. Initially developed as a means of monitoring expenditure on food, they evolved into surveys monitoring food intake and dietary trends. They are considered to be essential for this purpose. It was recognized that there are varying degrees of perfection for such surveys but that, for a particular one, the element of continuity was the most important aspect.

Also important was the possibility afforded to break down the results of these surveys and obtain information about specific groups of the population.

It would be desirable that surveys of different countries or regions were comparable but, on the other hand, account should be taken of financial and human resources constraints.

Specific surveys were mentioned (UK, Germany, The Netherlands, France). Over a long period, these identified a decrease in the consumption of potatoes, pulses and cereals and an increase in the consumption of meat, fruits and vegetables.

Over a more recent shorter period, a decrease was noted for energy intake, fat consumption in absolute terms but not necessarily in terms of percentage energy, and an increase in polyunsaturated fatty acids. Milk was a particular case.

Another notable piece of information was the variation of dietary habits within areas of the same region or area of a country and between income groups.

Apart from food intake, other surveys identified a number of interesting issues, e.g. that the concerns for consumers in order of priorities were different from those of the experts, and that the percentage of the population whose dietary habits were in conformity with recommendations varied greatly for individual nutrients (20% for fat, 60–70% for mono and disaccharides and 60% for cholesterol), with only 0.7% of the population following recommendations as a whole.

## SETTING OF DIETARY GOALS

The second point discussed was whether there is the necessity to fix dietary goals. What should these be? Intended for whom? What use should be made of them?

In this matter, summarizing the discussion has been a difficult task; but it would be said that some dietary goals are considered necessary. These can be a mixture of quantitative and qualitative ones. To whom these sets of dietary goals are addressed will determine their nature and degree of precision. It was generally felt that some quantitative goals could be addressed to the scientific community, which would have the responsibility of translating them more to qualitative ones for the average consumer.

## CONSUMER INFORMATION AND EDUCATION

It was felt that the two subjects, information and education, were closely related and that there should be parallel action in the two areas.

In the context of consumer information there was a short presentation of the recently adopted directive on nutrition labelling. It was felt that this directive represented a good consensus on what information was useful to be given to the consumer at this point of time. It was also felt that the information to be provided on the label would serve as a stimulus for the consumer to learn more about nutrition.

As was mentioned earlier, there were disparities between dietary habits/behaviour between different regions of one country and between different income groups of population, even though messages in the form of information or education were similar in all areas. This indicated some problems with reception or perception of messages. It was felt therefore, that messages should be more targeted to individual groups.

The decrease of overall milk consumption, despite an increase in low-fat milk consumption, would indicate that there has been a failure to transmit the correct message, while the depressing figure of 0.7% of people complying with the whole of the dietary recommendation would support the call made by Dr. Gibney to bring the behavioural sciences more into the scene.

## NUTRITION RESEARCH

The subject of research in the area of nutrition was not discussed exhaustively. However it was recognized that there exist a lot of gaps in our knowledge on nutrition which should be illuminated through research. The consequence of this situation is that prudence is necessary as to the precision of dietary guidelines issued.

The areas of nutrition research which were mentioned were:

— the relationship between diet and disease and the prevention of disease through better nutrition;
— the identification of markers of nutritional status.

The view was also expressed that, in the future, research relating to the area of 'nutrition and the elderly' will become important.

## FUTURE FOOD LEGISLATION IN EUROPE

Professor Rey presented his views on future food legislation in the community. He felt that:

— vertical harmonization should be the least possible, taking of course into account the projected specific directives in the area of foods for particular nutritional uses;
— more emphasis should be given to consumer information through labelling;
— Member States should play an important role in setting the programme and priorities of future legislation.

Two specific areas were mentioned for legislation:

— medical foods where the possibility of introducing availability of foods on prescription should be examined;
— food supplements and fortification or restauration of foodstuffs.

He stressed the importance of cooperation between the Commission and the national authorities, especially in providing scientific and other data, and the necessity for industry to make available well documented and presented files on all new substances and for the experts to examine these in an atmosphere of mutual respect.

# Part III
## Food science and technology

# III.1

## Trends in the perception of food quality by the consumer — his expectations and needs

**A. Huyghebaert**
University of Ghent, Faculty of Agricultural Sciences, Laboratory of Food
Technology, Chemistry and Microbiology, Coupure L. 653, B-9000 Ghent,
Belgium

### INTRODUCTION

In recent decades, profound changes have occurred in the quality perception of
foodstuffs by consumers.

In an agricultural society, consumers were very familiar with the production
conditions of foods and with the technology applied. In our modern society, the
distance between production and consumption has become very significant. Foods
are very often produced in foreign countries a long time before consumption. The
impact of preservation and transformation processes on foods is very great. Consumers, at the end of the food chain, have a number of quality expectations. More
and more questions are being asked about the overall quality of our daily food. The
attitude of consumers towards food is not only determined by technological factors
but also, and perhaps more, by socio-economic aspects.

This discussion intends to focus on the relationship between food and quality.
The consumer's standpoint will be confronted with the scientific approach. It is not
intended to give an overview of scientific and technological realizations in the past
and under present conditions. Instead, the tolerance to these technological achievements will be discussed from the point of view of the quality expectations of
consumers.

Quality covers several quality attributes: safety, nutritive value, sensory quality,
convenience and, finally, emotional value.

### SAFETY

#### Consumers vs scientists

No doubt consumers expect their food to be safe. Recent examples have demonstrated clearly that a public debate about a contaminant or a pathogenic microorganism has a tremendous effect on the acceptability of particular foodstuffs. These

overreactions are mostly temporary but have a long-term effect in that they contribute to the general feeling that our food is unsafe. *Listeria* in particular cheeses, *Salmonella* in eggs, hormones in meat, nitrates in vegetables and alar in apples, are not only topics to be discussed by the scientific community: they are very often part of the headlines in the news.

From the research, it could be concluded that safety is the most important quality factor of foodstuffs. It is the *sine qua non* for the acceptability of the product. Consumers always refer to contaminants such as pesticide residues, heavy metals, nitrates and others. There is more concern about potential toxic compounds of a chemical nature than about the really pathogenic microorganisms. The observation that there is a strong divergence between the public appraisal of food safety and the scientific evidence is in agreement with results obtained in other countries. Consumers are convinced that chemicals in foods are the greatest threat to their health.

Scientists do take into consideration the health dangers of chemical residues. There is, however, a large majority to confirm that the major health risks from foods are to be found in unbalanced nutrition and in faults in hygienic practices or microbial contaminants. Thus, the classification of risks in the opinion of consumers and of scientists is different. This fact illustrates that chemophobia has penetrated all aspects of our daily life.

One of the most important reasons for such misunderstanding is the concept of a calculated risk. Consumers judge the safety of foods in qualitative and not in quantitative terms. In their opinion a substance is toxic or non-toxic. The only possible scientific approach is to take the quantity of a compound into consideration.

A public debate on contaminants has become extremely difficult if not impossible. Misunderstanding of the problem is very common, as is clearly demonstrated by the famous case of E numbers and the Villejuif list. This example of misinformation is still widely found in groups dealing with the defence of consumer interests. After several years of efforts by different authorities to demonstrate the unsoundness of this list, consumers seem not at all convinced.

### Aflatoxins

Aflatoxins are another good example of a gap in understanding. These substances are secreted by particular molds. There is no doubt about the toxicity of some of these mycotoxins in general, and aflatoxins in particular. Until recently, aflatoxin levels were discussed in the ppb range, but nowadays in the ppt range. This is the result of progress in analytical methods to assess these very toxic compounds. Consumers, however, are very confused by this terminology. The dimensions of ppm or mg/kg, ppb or µg/kg and ppt or ng/kg are rather difficult to assess. The way in which data are communicated plays a very important role. An aflatoxin level of 10 ppt gives rise to a lot of anxiety whereas 0.01 ppb is less alarming and 0.00001 ppm is of no concern.

The communication gap between scientists and consumers is a basic problem. There is a need for more methodology in order to translate scientific data into a language understandable for the average consumer. Scientists are experts in their own fields but not in the communication sciences. There is room for initiatives by the scientific community in order to give objective information to the mass-media.

## Additives and contaminants

Confusion between additives and contaminants is very often observed. It is rather difficult for consumers to differentiate between:

— a substance added intentionally with the objective of fulfilling a particular function in food;
— a substance not intentionally added, but present and whose presence is undesirable.

Even after a full explanation, the final question is always, 'Is it really necessary to add chemicals to foods?'!

In a research project on the above-mentioned confusion, it was demonstrated that 65% of consumers are unable to differentiate between an additive and a contaminant. Only 35% are able to give a rough description of an additive, and a large majority of this group give a wrong definition in that they consider all additives to be toxic.

## Naturally occurring toxic compounds

Naturally present undesirable substances such as haemagglutinines, goitrogens and alcaloids, are a matter of less concern to the public. It is rather difficult for consumers to understand that potentially toxic compounds of natural origin are present in foods. Human beings have profound confidence in nature, compared with rather limited confidence in the result of their own work. It is evident, however, that nature is not so friendly as is generally thought.

## NUTRITIVE VALUE

### Preservation and nutrient retention

One major remark by consumers about the nutritive value of foods is that important losses occur during preservation, transport and distribution. It is rather difficult to understand how deterioration processes can be inhibited or delayed without damaging vitamins and other nutrients. A usual question is about the feasibility of inactivating living organisms or microorganisms without destroying the other vital elements in foodstuffs.

In scientific terms it is quite easily demonstrated that normal food processing operations result in highly nutritious foodstuffs. Optimalization of heating processes, such as the high-temperature short-time principle, can be theoretically explained and practically proven. However, there is again a communication gap. Explaining these phenomena to consumers who are without a basic knowledge of reaction kinetics, heat transfer, process and product technology is rather difficult. It has to be translated into simple terms that foods stabilized by preservation techniques show an adequate nutrient content. Consumers have to be convinced that there is no contradiction between preservation and nutrient retention.

### Transformation processess

With respect to transformation processes, profound changes have taken place and important progress has been realized. Transformation processes are defined as those processes where separation and recombining techniques are applied. One of the major development areas in food technology is the separation of raw materials into proteins, fats, carbohydrates and other substances, followed by the modification of the functionality of these ingredients and their recombination into consumer products.

One major advantage of these developments is the possibility of producing foods adapted to specific needs: low fat products, products with low salt and sugar, high fibre and low cholesterol content, and products with a modified fatty acid profile. As a result, consumers are nowadays exposed to a range of foods with particular nutritive properties. The message of these products is only partly or not at all understood. There is real risk of misinterpretation.

The yellow fat market is a typical case of non-transparency. Products are available with a high, medium and low fat content, with a variable fat composition: milk fat, non-milk fats, vegetable fats and blends. There is a lot of confusion about yellow fats as a lot of labelling is misunderstood.

It has to be emphasized that positive developments with respect to food composition have to be presented to consumers in such a way that they are identifiable.

## SENSORY PROPERTIES

### Selection criteria
Flavour, texture and appearance are essential quality attributes for the acceptability of foodstuffs. From our own research it could be concluded that, once safety conditions such as the *sine qua non* are respected, foods are mainly selected according to sensory properties.

In the first place, criteria like freshness are of utmost importance. Freshness is even more important than nutritive value and other criteria for the choice of a particular food.

Every effort towards a better nutritive profile of products, by an adaptation of the composition, has to take into consideration the major importance of sensory quality. Very often a reduction in fat or salt content or an increase in fibre content results in a lower sensory profile. Novel production and transformation systems and new products are only acceptable for consumers if basic demands for flavour, texture and appearance are respected.

### Freshness
Freshness has become a magic word. Freshness is used by consumers in a broad sense and includes also the natural character of the product. Scientifically it is rather difficult to define exactly what a fresh product is. There have been many attempts to give a definition.

Is the criterion the harvesting or the production period? What technologies can be applied that respect the freshness property? What are the limits beyond which a product can no longer be considered as fresh? There is a need for a clear definition of fresh foodstuffs.

## The 'natural food' trend

The trend towards natural foods presents particular risks. Pasteurization processes are traditionally applied for dairy products with the objective of producing safe foods. As a result of adequate heating processes, transfer of potentially present pathogenic microorganisms to human beings is not possible. This is one of the basic concepts in food technology.

From a scientific point of view, developments towards natural foods are a subject of much concern. There is indeed for these products a tendency to market dairy products manufactured from raw milk, meat products without nitrite or nitrate, minimally canned vegetables. The idea behind these developments is to produce minimally processed foods without additives or technical aids. Chemophobia may result in increased problems with pathogenic microorganisms.

## The food technologist's dilemma

Food technologists are confronted with opposite demands: foods with both a fresh character and a good shelf-life. There is a continuous search to solve this dilemma. It is clear that there are possibilities for novel technologies. With respect to consumer attributes, physical processes are better accepted than the use of chemicals. Examples are membrane techniques for the elimination of microorganisms and supercritical extraction as a soft separation technique.

## CONVENIENCE FOODS

### Recent developments

There are many examples of convenience foods that have been developed during recent years: ready-to-eat meals, fourth gamma vegetables, microwaveable foods, vacuum-packed cooked meals. There is a trend towards further convenience. More and more food preparation techniques are being taken over by the industry. Consumers are no longer prepared to spend much of their free time in the preparation of meals. Demographic and social factors play an important role in this development. It is however striking that consumers are prepared to accept convenience food quite so easily. Subsectors of the food industry have recently developed in this area. The potato processing industry has taken over the preparation of potato products, including the frying of french fries. There has been a long development from the situation of own garden production of potatoes to the present-day situation of ready-to-eat potato products.

### Quality factors

The penetration of convenience in the food system has important consequences. The responsibility for quality factors like safety, nutritive balance and sensory properties is transferred to the industry. Part of the identification with particular foods is lost. There is a trend towards more uniform foods. For some people this evolution is a matter of concern because it is felt as a decrease in the quality of life.

## THE EMOTIONAL VALUE

### Factors in product image

The quality of foods includes more than the quality attributes so far discussed. There is an emotional value which is an important factor in the quality perception by consumers. The image of a product is the result of several factors and is continuously changing. The emotional value is to a large extent determined by the value attributed by the society to particular processes. It is related to:

— the impact on the environment of processes in agriculture, industry and distribution,
— the energy consumption and the production of waste,
— the living conditions of production animals,
— the use of fertilizers, pesticides and other production aids.

It has been demonstrated that neither the nutritive value nor the sensory quality of free range eggs is higher than of normal eggs. However, a section of the consumers is definitely interested in these eggs and is prepared to pay an additional price for them. Similar comments can be made for pork and about organic vegetables.

### Consumer tolerance to technology

Consumers of the end of the twentieth century are interested in the way foods are produced, transformed and distributed. Recent discussions about the quality of tomatoes according to the production system, like hydroponics and artificial substrates, confirm this observation.

There is a well defined tendency towards more traditional production processes. Statements by scientists that production aids have no negative effect on a product seem to be of secondary importance. To a large extent consumers are convinced that the use of production aids results in imbalances in the composition and the overall quality of foods.

It will be rather difficult to convince consumers of the merits of a substance like BST (bovine somatotropine). Arguments of a socio-economic and toxicological nature will not be sufficient.

The evolution in attitudes towards foods has as a result that other values have to be taken into consideration. The debate is no longer limited to scientific matters. A new process or a product will be scrutinized from different points of view.

There is evidence that tolerance to technology will be increasingly important for the acceptability of foods.

## CONCLUSION

There are differences in quality perception by consumers and by scientists. Consumers are interested in the way food is produced, transformed and distributed. There is a real need for a better communication between consumers and the different parties involved in the food system.

In order to re-establish confidence in foodstuffs, the food system, in all its aspects, needs more transparency. Quality of foods is important as it is an integral part of quality of life. Scientists are convinced of the high quality of foodstuffs in our western society. The challenge is to bring the message to the consumer.

# III.2

# Opportunities offered to product development by raw material and food ingredient innovations

Peter de Vogel
Research and Development Director (Brewing), BSN Group, 7 rue de Téhéran, 75008 Paris

The title of this chapter requires some preliminary explanation in order to make it compatible with the ideas which will be presented here. Innovation is not merely a technical achievement which the scientist or technologist may regard as a breakthrough which meets his objective. In order to become truly innovative, the technical achievement must be accepted by society, it must meet the socioeconomical criteria so as to represent a new added value. Hence it might have been more adequate to rephrase the title to address the question as to how new raw materials and food ingredients may have an impact on food product innovation. People buy and consume food, not ingredients nor raw materials, at least not in the industrial meaning of these terms.

The historical record of innovation in food products can be appreciated in two distinctly opposite ways. The pessimistic view could be that nothing has basically changed. Mankind needs food for its survival and that food is still made up of, essentially carbohydrates, proteins and fats with some minerals and vitamins. But there is a view which is more exciting: never has mankind had access to such a wide array of products from which he may choose in order to satisfy his needs in terms of health criteria, sensory satisfaction, physiological needs or simply economy or convenience.

Clearly one can only speak about innovation with the second view at heart. The food product manufacturer is the penultimate participant in a chain of techno-economical events. This chain comprises breeders of seeds and livestock, agriculture, animal feed, agro-chemicals, raw material harvest and storage, primary and possibly secondary transformation, food product manufacture, food packing, preservation, storage and finally distribution. All these factors can contribute with a technical or economic factor to the final innovative event which in a consumer product market requires consumer acceptance.

Before making an attempt towards a prospective view, let us first look back at the history of the food industry and its innovative records.

There is a general feeling that food product manufacture requires good craftmanship rather than ingenious 'high-tech'. This is the consequence of the sharp contrast between this industry and, for example, the communication or spacecraft industry. The breathtaking results of telecommunication, television and computer technology have led to entirely new markets and even to totally new life styles. These innovations stem, from breakthroughs like, for example, semi-conductor technology and transistors, which in turn are the result of the ever-increasing knowledge in material science. Similarly, the rapid progress in medical care is the result of a large variety of scientific developments, not only in chemistry and biology but also in diagnostic devices again making use of electronics. It should be noted that most of these high-tech industries spend between 10% and 20% of their annual turn-over on research and development, while in the food industry such expenditure on an average hardly reaches 1%. Most food products have a proportionally lower added value and hence profit margins in such a fragmented industry are relatively low.

It is interesting to observe the trend in expenditure on food products as a percentage of the family budget over a period of 150 years. Until early in this century, for most of the population, procuring food in order to survive required the major part of their financial resources. Only since World War II has the figure dropped significantly below the 50% level (Table 1). Nowadays the percentage in Europe lies around 18% while in the US it has fallen as low as 14%. This does not mean that people spend less on food in absolute value, but its cost has not kept pace with inflation and certainly not with the average real income.

Another economic trend over this long period is the sharp increase in the consumption of food products which have been subject to some form of transformation. The industrialization phase was followed by the development of marketing and distribution strategies which have led to the present situation, where more than 80% of the food consumed is sold under a brand name or at least carries the label of a packer or distributor. Consequently the consumption of commodity products and bulk foods has dropped significantly (Table 2).

The pasta industry offers a striking example of how innovation can bring about such trends. Pasta has traditionally been a bulk food made from cereals either in the home or in the artisan foodshop or restaurant. Large-scale industrial production offered the possibility of extended keepability; as a consequence, pasta became one of the products to be stored in case of food scarcity. Its main function, as a source of energy-rich food, led to its having a poor image in the market, and consequently the consumption of the industrial products dropped after the last World War.

More recently, better understanding of the raw materials and processes has enabled the industry to produce branded high quality products and to diversify the presentation by the use of a range of new ingredients, varying colour, flavour and texture. Nutritional research has shown the value of pasta as a source of energy which is not stored so as to produce overweight, but which is perfectly balanced to physical effort. Thus pasta has achieved an increasingly favourable image. Its consumption is

**Table 1** — Evolution of household expenditure on food as a percentage of the family budget (France)

| Year | % |
|------|------|
| 1830 | 90 |
| 1950 | 50 |
| 1970 | 24.5 |
| 1989 | 18 |

**Table 2** — Evolution of the average annual per capita consumption of commodity foodstuffs (in kg)

|  | 1965 | 1985 |
|------|------|------|
| Bread | 80 | 46 |
| Potatoes | 95 | 40 |
| Fresh vegetables | 71 | 60 |
| Meat | 21 | 22 |
| Fish | 6 | 6 |
| Milk | 84 | 66 |
| Sugar | 20 | 11 |

steadily growing and is now spreading from the southern European countries towards the north.

Just as much as techno-economic opportunities push innovation, socio-economic trends create opportunities for product innovation. The growth of collective restauration has led to 'fast food', which is only in its early stages of technological achievement.

The decrease in time available for home cooking has opened up the way for deep-frozen, ready-made, microwaveable foods. The phenomenon of irregular food intake has given rise to a huge snacks market. The desire for diversity to satisfy the palate, satisfying which used to be the role of the housewife, has moved the market to a large variety of portioned food, often of an exotic nature. Finally, growing health awareness by the consumer has created a new potential for innovation. Low calorie food, low cholesterol diets and high fibre foods are leading towards a new market of functional food products (to which we shall return briefly below). In the brewing industry, alcohol-free beers have existed for several decades, but it is only in recent years that these markets have had an annual growth of more than 10%.

All these trends will increase the need for new food ingredients and specially designed raw materials, which meet very strict specifications. New processes will have to be designed to prepare the ingredients and to obtain the desired end-product. Packaging and preservation technology will have to contribute to the development of new and safe distribution systems.

The opportunities offered to the food industry by new ingredients and raw materials also become obvious when one considers the history of the food industry.

In the past century, most food products consisted of farm produce which was sold direct to the consumer. This fact has set the scene for existing food habits in the various areas of Europe. These habits still exist and are so profoundly anchored that they constitute an almost inborn reference for the consumer. However impressive the changes may seem, the evolution is slow and every now and then trends revert to the deeply felt security of the original sources. Who does not once in a while prefer his home-grown vegetables, or choose to buy a fresh cheese direct from the farm?

The first stages in the development of the food industry were the consequence of general industrialization stemming from the opportunities offered by mechanical processes (milling), vapour and refrigeration and transportation. Food produce thus became available over a wider geographical area and could be sold in the rapidly growing urban areas.

The first technologies resulting from this phase were canning and later dehydration, applied mainly to agricultural produce. Strict rules regarding practices were required to ensure public health. Stringent regulatory actions were necessary to avoid food adulteration and fraudulent practices. Many of these early rules, which went into great detail, are still in force today. They cover food preservation techniques and quality criteria in terms of composition and acceptability of recipes.

Demographic growth and wider geographical distribution, required for example for troops on the battlefield, called for increased shelf-life. Increased knowledge of the chemistry and microbiology of food products led to the introduction of ingredients which were not nutrients in their own right. Research into ancient preservation practices, such as salting, drying and fermenting, led to a new category of ingredients called food additives. In terms of flavour, colour, texture and preservation these ingredients allowed the industrialization of food products to progress.

Although there is no record of serious hazards due to the use of food additives, the consumer became suspicious and health authorities started investigations. Hundreds of millions of dollars have been spent over the years to establish the safety of food additives, and regulatory action has been taken to draw up 'positive lists'. Levels and conditions of use have been established and there must exist a technological purpose for their use. Moreover, in most countries in the world the presence of additives in the food has to be clearly stated on the label. It is therefore disturbing to see that the consumer still displays a distinct aversion for these safe ingredients. There is a matter for thought in deciding how to cope with future new ingredients and raw materials from a regulatory point of view.

Over the last twenty years, a large variety of food ingredients obtained by fractionation of agricultural produce have found their application in food product innovation. Fractionation may be followed by hydrolysis (enzymatic and chemical) or even by chemical transformation. Fractions and derivatives from milk proteins (in particular casein), from carbohydrates (in particular starch) and from fatty acids and triglycerides have enabled industry to break away from traditional recipes and to introduce new processes. Flavour compounds form a very large separate category of ingredients which have become essential in the creation of new product concepts.

Industrial processes which originally were adapted to traditional recipes and which therefore carried numerous constraints, which do not allow, for example, for productivity increase or for quality improvement, can now be entirely redesigned with these new ingredients.

Such ingredients do not create the same degree of suspicion with the consumer as additives, as long as they originate from traditional agricultural produce like cereals, milk or other vegetable or animal sources. Things become different when such ingredients are obtained from non-traditional sources, as for example protein from unicellular organisms. Ingredients are then often referred to as 'substitutes'. Such terms may carry within themselves qualifications which put innovation at risk.

Once a new ingredient, whatever its origin may be, meets the requirements of toxicological and nutritional acceptability, there should be no additional constraints to its development in the market.

Although the chemist and the toxicologist may be satisfied with the adequate specification and the safety of a new ingredient, two new problems arise. The mixing of ingredients and their possible subsequent processing may lead, just as much as in traditional food manufacture, to the formation of substances which either are not safe or which modify the nutritional value of the food. It is therefore important that the assessment of food safety if always applied to the end-product. The second point of concern lies in the fact that broad variations in food composition and changes in eating habits may lead to pronounced changes in the overall intake and may thus lead to nutritional imbalance. The use of soy-protein in the development of dairy-type products could lead to a decrease in calcium intake. The consequence might be that calcium supplementation of such products is required.

Another example where eating habits are important is dietary fibre. It is generally believed that modern eating habits have reduced the intake of dietary fibre. While extensive research is being carried out to define chemically such compounds and to determine their exact role in nutrition, the industry has reacted by supplementing food products with raw material fractions believed to be rich in fibre. If supplementation is applied to foodstuffs which are recognized by the consumer as being potentially rich in fibre, like bread, there is no risk. One can even add fibre to foodstuffs which are eaten in combination with cereal products like dairy products. But is it reasonable to add fibre to foods which never have been expected to be a source of fibre, like some mineral waters in the Far East?

The consumer is informed by adequate labelling about the composition of the products, but it is not clear whether this information reaches him and what he does with it.

The wholesomeness of a foodstuff is the responsibility of the producer. Regulatory bodies cannot hope to regulate each individual novelty stemming from ingredients or processing. Hence modern regulatory action should focus on the obligation of the assessment of wholesomeness by the manufacturer. It is up to the industry to take the responsibility for wholesomeness as an integrated part of product innovation.

The most recent step towards modern food manufacture is represented by the potential opportunities offered by the genetic control of vegetable, animal and microbial life. Most of the traditional food processes have been designed to obtain the desired product from available agricultural produce. This meant that only an increasingly strict selection of raw material could allow the improvement of the quality of the end-product. The use of so-called technological aids in food processing occasionally allowed improvement in productivity.

The recent techniques in breeding offer the challenging potential for tailor-made raw materials. The opportunities which may be foreseen are in most cases still of a rather hypothetical nature. Indeed, in order to bring about modifications at a molecular biology level, a full understanding of the metabolic pathways and the corresponding genetic regulation systems is required. It is unlikely that practical developments in this area will be of a revolutionary nature. The complexity of the process of bringing about a specific genetic modification in plants or animals is still such, that the cost will only in specific cases be offset by a sufficient economic advantage. The increasing regulatory constraints in this field of research have put an additional limit to rapid progress in this area.

Yet the opportunities are manifold. An increase in added value in agriculture may be expected. Better quality and possibly lower production costs and higher yields may be achieved in transformation processes. Plant disease may be controlled by means other than the use of chemicals. The levels of natural toxicants present in traditional food raw materials may be significantly reduced, and finally the presence of desirable traits relating to sensory quality or nutritional value may be increased. It may therefore be expected that the introduction of these new technologies will occur in an evolutionary way and certainly without any risk to the consumer.

To conclude this prospective view of raw materials and ingredients, we must return to a more recent class of products known as functional foods. Raw materials used in food manufacturing are made up of a vast variety of chemical compounds, many of which have a physiological or pharamacological activity. As long as such components have not been identified, their presence is at best experienced as a beneficial effect of a complex but traditional product. But, once these compounds have been identified and can be separated from the raw material and purified, they become ingredients. Such ingredients, chemically identified and the action of which is well established, may then be used in the creation of functional foods. The growing health awareness of the public and the increased understanding of the role which food plays in general well-being and in the prevention of nutrition-related diseases, could lead to a rapid growth of such markets.

General functional objectives are related to overall disease resistance, for example by non-specific immune response. Obesity prevention, anti-ageing effects and prevention of osteoporosis are other examples where functional foods may present an interest.

A large number of food components are under investigation in order to establish their exact chemical identity, their pharmacological action, their dose–response relationship and the optimal ways for their introduction into the diet. Oligosaccharides, peptides, essential fatty acids, natural antioxidants, free-radical scavengers, natural bacteriostats and immune response activating substances are on the long list of potential future food ingredients.

Such developments will move food products of the future closer to the health care area. But despite the research effort required to substantiate their physiological effects, functional foods should, above all, remain food. It should be freely available and therefore it should be possible to consume it at levels beyond the optimal range, without any risk to the consumer. Functional foods must therefore be just as safe as traditional food. Any specific claims which may be made regarding their effects on health must be well documented and the consumer must be adequately informed and educated. All this is an unseparable part of the innovative process.

Innovation, particularly in the food industry, is in a way 'the management of contradictions'. It is the act of creating a new combination of elements leading to a new set of values which will ensure its socio-economic integration into the market.

The contradictions in consumer attitudes which we constantly face can be described as follows:

(1) The consumer accepts high quality at an increased cost, BUT spends less money on food.
(2) The consumer favours 'traditional recipes', BUT is eager to experience exotic food.
(3) The consumer favours the possibility of a wide choice in the market, BUT he is often unable to constitute a balanced diet.
(4) The consumer is eager for new experience, BUT rejects certain technologies used to achieve the product innovation.
(5) The consumer has increased his health awareness, BUT often fails to balance his food intake between pleasure and need.

Basically this means that our products must always meet consumer acceptance, and in order to achieve this we need to investigate consumer attitudes to products. The question is much more, 'What does the consumer do with a product?' rather than, 'What do we think he ought to do with it?'.

Secondly we need to improve the education of consumers on matters regarding the food we offer them. In the modern food market the tradtional cultural items, which were the reference for the consumer, have been removed. Despite extensive informative labelling he is unable to establish a balanced diet.

The basic principles are that the consumer is perfectly able to judge on matters regarding sensory acceptance and convenience; but he has only a very fragmented understanding of the relationship between his health expectations and the food he eats. And last but not least, he wants to be reassured on food safety.

In conclusion, scientific progress is opening up vast possibilities for product innovation, in particular by means of raw materials and ingredients, but there is no reason to expect a revolution as long as the consumer does not want it. Food scientists and technologists are no wizards; the consumer must come to understand that they behave in a responsible way.

# III.3

# New process opportunities for product development

**Joseph Lenges**
Director of the Experimental and Analytical Station of CERIA (Centre for
Education and Research for Food and Chemical Industries), 1, Avenue Emile
Gryzon, B-1070 Brussels, Belgium

## INTRODUCTION

For several years we have been witnessing a profound restructuring and international regrouping a food manufacturers in the same sector with a view to increasing their competitiveness and commercial potential at the approach of a free market for both goods and men. This change from agro-food businesses into multinational units favourises innovation and the emergence of new techniques through the implementation of adequate structures and through the application of food engineering procedures. This evolution is stimulated by communication and collaboration among the different physico-chemical sciences (rheology, colloidal chemistry, organic and analytic chemistry), the engineering sciences (food engineering, process control and automation) and the socio-economic sciences; the last named make available to us the tools to perceive the motivations and aspirations of the consumer.

The objective for any food industry is the manufacture of quality foods which will satisfy both nutritional and taste needs within a perspective of profitability and respect for man and his environment.

The 'quality' aspect is fundamental. It constrains the food industry to certain obligations which necessitate that the concept of 'quality' should be conceived, taking into account not only composition specifications as regards nutritional and legal demands, but also subjective obligations, conditioned by organoleptic properties and the behaviour of consumers, and hygiene constraints, which guarantee the consumer a food product free from any contamination.

Innovation in the food industry consists in bringing to it a positive contribution to the concept of 'quality'. This development can be realized by an improvement of quality in any number of ways, by a reduction in the price of production (investment and operating costs), by an improvement in the tasks of the operators and by a reduction in smell and hearing stress. It is the engineer responsible for a process who must assemble and evaluate any constraints, as well as note his objectives towards progress and maintain a global vision before changing his decisions.

These objectives can be reached through the methodology of food engineering and greatly helped by computerization and automation. The conception of new units of production is realized by the quantification of mass flow and by the establishment of sequences of unit operations. Once mastered, the whole procedure is put into operation in the form of expert systems, which enable the collection of information, its treatment and the taking of adequate and immediate decisions without man but within a logical framework which he will have established beforehand.

The role of the engineer will be to make the system flexible and reliable with a view for guaranteeing the manufacture of a product of constant quality responding to a product description established together with the marketing department.

In the domain of the food industry, chemical engineering together with food technology has given birth to food engineering.

Given the importance of techniques of food preservation, food engineering has fixed, as its principle objective, the study of unit operations of a physical nature. The application of chemical engineering methods to the study of biological reactors has given birth to biochemical engineering, which includes microbiological engineering whenever quantifying the design of reactors with microorganisms is involved, and enzymatic engineering, when the operation is limited to the use of enzymes. As a body, these disciplines, based on the concept of chemical engineering, are today grouped under the general name of food process engineering.

## MASS FLOW

Mass flow in the food industry can be illustrated as shown in Fig. 1.

The role of the food industry consists of transforming agricultural raw materials by submitting them to different preservation and transformation techniques and by conditioning the final products in appropriate packaging. The use of these different techniques takes into account the special nature of the raw materials and the importance of the organoleptic and dietetic properties of the finished foods.

## TECHNIQUES USED

Obtaining new products through industrial treatment implies the use of one or several of the techniques illustrated in Fig. 1.

### Preservation techniques

These increase the shelf-life of the products and consequently enlarge opportunities for storage in time and space. They are applied to prevent alterations of a microbial, enzymatic or purely chemical nature. The preservation techniques which prevent microbial development are especially treatments by heat (pasteurization and sterilization), by cold (chilling and freezing), by reducing water activity in the products

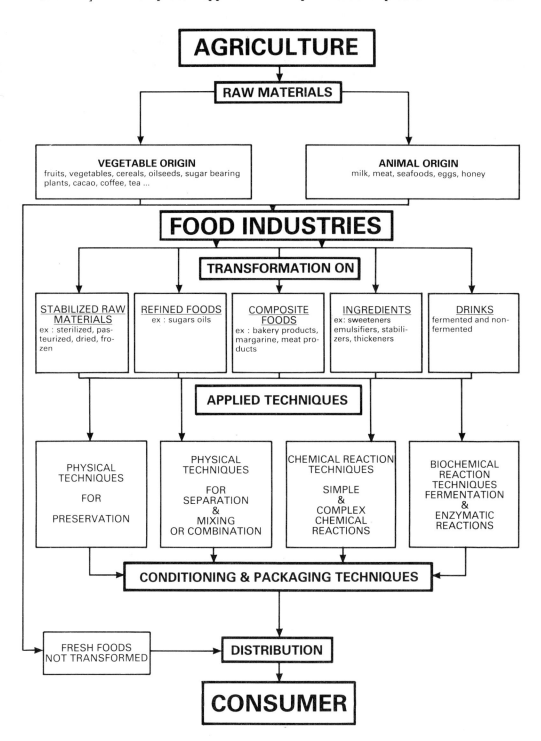

Fig. 1.

(addition of sugar or salt, drying), by ionizing irradiation, by sterilizing filtration, by acidification (lowering of pH) and by the addition of antiseptics.

The inhibition of enzymatic reactions can be accomplished through treatments analogous to those applied for the inhibition of microbial alterations.

Preservation techniques against chemical alteration are chosen according to the substrate and the process of alteration. The phenomenon of oxidation can be combatted by vacuum packaging or packaging under inert gas or by the addition of reducing agents or by the limitation of oxidation catalysts during the production process. Non-enzymatic browning reactions can be avoided by means of:

(a) the elimination of traces of sugars present in certain foods — this is especially the case in the treatment of eggs with glucose-oxidase prior to drying;
(b) by lowering the pH;
(c) complete desiccation, as far as water activity level lower than 0.5; and finally,
(d) by the addition of inhibiting substances.

It is important to be extremely prudent when putting these last-mentioned treatments to use.

### Production techniques by separation and mixing of phases

New products are characterized by a composition and by organoleptic properties which are very different from those of the raw materials. Formulation of the new product necessitates mastery of the functional properties that govern the behaviour of the product during the production process, but equally, a careful choice of which unit operations or techniques to use.

Among the separation of phases techniques, the engineer has the following procedures at his disposal: mechanical extraction or extraction by solvent, sedimentation or centrifugation, filtration, microfiltration, ultrafiltration, reverse osmosis, evaporation, distillation and drying. We can also include the mechanical preparation of products by milling and sieving, which also constitutes an operation of separation.

The techniques of separation are especially applied in milling, in starch manufacture, in the oil, butter and sugar industries and in the manufacture of decaffeinated coffee and gelatine.

Mixing or combination techniques are scrupulously perfected in the manufacture of margarines, mayonnaises, ice-creams, pastes, spreads, sauces and jams. We can also include chocolate and confectionary manufacture and most food specialities.

During separation and mixing operations there often take place granulometric, rheologic and organoleptic modifications as well as transfers between phases.

As examples of modern techniques we can quote the agglomeration or instantizing of particles (instant powders), the encapsulation of aromas and hygroscopic products and the formation of water-resistant barriers in composite food products.

### Chemical reaction techniques

Simple chemical reactions are far from being as important in the food industry as they are in the chemical industry. They are applied principally in the fats industry.

In general, the heat treatment of a biological product entails complex reactions which are responsible for favourable or unfavourable modifications in taste, colour and smell. These are most evident in torrefaction, baking and cooking.

### Biochemical reaction techniques

Biotechnology in the food industry exploits the potential of microorganisms and enzymes in order to modify the properties of foods. The most ancient application of microorganisms is surely fermentation; in this case, enzymes and microorganisms act directly on raw materials, modifying them and restoring them into food.

Amongst the main transformations of this kind we can list the manufacture of beer, wine, cheese, yoghourt and sauerkraut. Today, starter cultures or selected microorganisms are incorporated into raw materials to effect precise biochemical transformations, generally linked to the phenomena of preservation of that food. Their use enables the mastery of a fermentation process which entails organoleptic modifications and improved appearance, which contribute to the quality and preservation of the products.

It is particularly in the dairy and meat industry sectors that the use of effective and specific strains can contribute to the establishment of high quality products which can be classe as 'new' products.

Bioconversion using enzymes constitutes another means of modification of raw materials; here we are thinking mainly about the application in the starch industry for the production of glucose and fructose syrups, and in the cheese industry for coagulating milk and for the hydrolysis of lactose and casein. These enzymatic reactions, compared with classical chemical reactions, are characterized by their specificity, in that they accomplish only one transformation within a complex field, and through a reactional medium which is little distorted, formed of water at very low temperatures.

### INNOVATIVE SECTORS IN THE FOOD INDUSTRY

The consumer wants products whose organoleptic properties are as close as possible to the fresh product and which contain few, if any, additives; and the consumer wants food products with a better suited nutritional quality, which is to say, containing less salt, less sugar, less saturated fat, more polyunsaturated fats and, if possible, a lower calorie level.

In this context, it is the preservation techniques which are most relevant. They are summarized in Fig. 2.

To innovate in this area will call for techniques entailing a minimal destruction of organoleptic qualities. It is pointless to submit to a taste panel a soup prepared with fresh vegetables and the same prepared with pre-dried vegetables; the one which gives the most satisfaction and pleasure to eat will be identified without any trouble.

The best way to preserve the original organoleptic qualities of foods is freezing. This technique applied at an optimal freezing speed ensures the consumer is guaranteed foodstuffs closest to the fresh product. However it is advisable to store

| Objectives | Parameters | Techniques |
|---|---|---|
| Slowing down or inhibition of microbial growth | Lowered temperature | Chill-storage |
| | | Freezing and frozen storage |
| | Reduced water activity | Drying |
| | | Curing and salting |
| | | Conserving with added sugars |
| | Decreased oxygen | Vacuum and nitrogen packaging |
| | Increased carbon dioxide | Carbon dioxide-enriched 'controlled atmosphere, storage or 'modified atmosphere' packaging |
| | Acidification | Addition of vinegar or other organic acids |
| | Alcoholic fermentation | Brewing, vinification |
| | Use of preservatives | Addition of preservatives (inorganic, organic) and antibiotics |
| | Nutrient restriction | Compartmentalization of aqueous phases in water-in-oil emulsions |
| Inactivation of microorganisms | Heating | Pasteurization |
| | | Sterilization |
| | Ionizing irradiation | Radurization |
| | | Radicidation |
| | | Radapperitization |
| Restriction of access of microorganisms to foods | Decontamination (ingredients and packaging materials) | Heat treatments |
| | | Irradiation |
| | | Disinfection |
| | Aseptic processing | Aseptic thermal processing and packaging |

Fig. 2 — Major food preservation techniques [11].

frozen food at a constant temperature around −18°C to −20°C so as to avoid the modification of ice crystals; these are particularly produced in non-structured foodstuffs when they are submitted to fluctuations in temperature.

The development of freezing is an innovation in the sense that it is advisable to research the optimal conditions for preservation.

The heat treatments of pasteurization and sterilization of products in bulk, followed by an aseptic conditioning, presents another alternative for the food industries. These treatments, applied at optimal temperatures and for short times, ensure, in a practical way, the preservation of organoleptic and original nutritional properties. This technology, introduced in the dairy industry in the 1960s, has now reached an advanced stage of development in Europe.

The application of this technique to viscous and heterogeneous liquids has been the subject of a feasibility study in our Centre, and seems to be of prime importance in the production units of large multinational food companies. It is advisable to define the best methods of heating in order to ensure the expected heat effect on the destruction of microorganisms and enzymes, whilst, at the same time, preserving to the maximum, the original organoleptic properties of the foodstuffs. The new ways of heating under study are heating by microwave and ohmic heating or heating by electric conduction [1, 2]. In the case of particulated foods in a liquid phase, it is advisable not to overheat that liquid phase in relation to the solid phase. The two methods of heating mentioned above seem, up to now, very promising, and lead to positive results for particulated foods. In the same context, the heating system for particulated foods called 'single-flow' FSTP (Fraction Specific Thermal Processing), developed by the STORK company, enables the same objective to be reached by controlling the distribution of the residence times of the two phases in the pasteurization or sterilization temperature holding section (STORK Rota-hold and STORK Spiral-hold) [3, 4].

This technology is applicable to all suspensions with a viscosity, at the liquid phase, from close to that of water (1 cP) up to the consistency of a pumpable double cream (700–1000 cP), with particles up to 25 mm in diameter and with concentrations of particles varying from 15% (cubic particles with sides of 18–20 mm) to 40% (spherical particles 6 mm in diameter).

The quantification approach to the heating procedure for particulated foods is much more sophisticated then that for simple liquids (milk, fruit juice and vegetables). In the case of suspensions, the evolution of the temperature in the centre of the largest particle and its time in the heating system must be taken into account. For this it is necessary to investigate the heat transfer between the phases, to measure the distributions of residence times in the different thermal treatment systems and to quantify the microbic populations before and after treatment.

Uperization, another sterilization technique in use in the dairy industry since the beginning of the 1970s, has undergone a revival [5]. Uperization is a procedure by which the transfer of indirect heat through a partition is replaced by direct transfer, consisting of injecting a quantity of steam into the liquid in order to reach sterilization temperature instantaneously. After holding the liquid at sterilization temperature, there follows cooling by expansion or vacuum cooling, which, at the same time, causes the evaporation of the quantity of steam previously injected. The range of liquid and homogenized foods treated this way includes milk, soft white cheeses, ice-cream mixtures, custards, egg products, juices, soups and sauces [6].

Recently, uperization was successfully tested on particulated foods in liquid phase; heating was realized by the injection of steam, and cooling, by the injection of sterilized and de-aerated water. Perhaps a combination of an appropriate heating technique and a selective technique to maintain the sterilization temperature of the two phases will prove to be the optimal solution for the sterilization of particulated foods at high temperatures over short times.

No other preservation technique has had as much press coverage as treatment by ionizing radiation. For 30 years this technique has entailed numerous toxicological studies, research which never seems to reach any end, so great are the demands for guarantees of security from the consumer and public authorities. Conclusions on all these studies were published in the form of recommendations by the OMS in November 1980 (Technical Report Series 659) and reconfirmed in 1987 (OMS N 40, Service des Media, *De point en point*) [7].

Food irradiation up to an average dose of 10 kGy or 1 Mrad presents no risk from the toxicological point of view; 95% of foods require an absorbed energy dose well below 10 kGy to be decontaminated [8].

Under good GMP conditions, the nutritional and microbiological aspects linked to irradiation are, on all points, comparable to those presented by the other conventional techniques of pasteurization or sterilization. Irradiation completes existing techniques, being applied especially to granular or powder products such as spices and thickeners, to tubers for desprouting and to products which are susceptible to insects and parasites. Products with a high hygiene risk (for example, small shellfish, frogs' legs) and pet food products should also be considered for this treatment.

Among the mixing and retexturing techniques which appreciably modify the physico-chemical properties of prepared foods, we should note, especially, the reconstruction of seafood products in the form of protein jellies (shellfish analogues: crab, shrimps and lobster tails) and the microparticulation of milk and/or egg proteins, which lead to structure and taste properties analogous to those of fats (Simplesse®, Trailblazer®) [9].

Every year, two million tons of fish are transformed into Surimi in Japan. These products are used as the raw materials for fish analogues. The technology for the preparation of Surimi is based on the extraction, by washing in water, of the flesh, of compounds of low molecular mass, fats, and soluble proteins, called sarcoplasms. Thus is obtained a concentrate of myofibrillar proteins to which is generally added 4% of sucrose and 4% of sorbitol before freezing. The addition of cryoprotectors enables the product to be preserved in its frozen state for more than 12 months without any loss of its jellifying qualities. After defrosting, salt is added (2–3%) and this mass is cooked to bring about the formation of a firm and elastic jellified network [10]. According to the type of analogue manufactured, starch, egg white, soybean proteins, oil, natural flavourings and colourings can be added.

Retexturing fish proteins offers the opportunity to innovate with regard to the appearance or presentation of the final product. The application of this retexturing technique under good hygiene conditions enables this branch of the food industry to conceive a whole series of products of a high nutritional quality.

The development of low calorie foods has caught the attention of numerous R & D laboratories. Excessive consumption of fats has stimulated the perfection of new products with a reduced calorie level of fats and dry ingredients. The first most prominant developments were minarine and 'light' butter.

The recent development of a fat substitute was realized with proteins from egg or milk and other raw materials as the starting point. The product, called Simplesse, was the first one commercialized at the beginning of 1991 as an ingredient in frozen desserts based on milk and was recognized as safe for human consumption (GRAS) by the FDA. Simplesse is produced by an internationally patented, heating and blending or shearing process that shapes the protein gel into microspheroidal particles which the tongue perceives as fluid rather than as individual particles. The sensory impression is, it would seem, as creamy and as full and rich as that associated with fats. This substitute generally contains two-thirds water, and consequently its calorie value only represents 15% compared with the same mass of fat. It is therefore an ideal substitute for fats. Permission for this product to be used in other non-frozen dairy products such as sauces and mayonnaises is being waited for.

Given that it is a protein jelly, it cannot be incorporated into foods to be heated, cooked or fried; a thermal treatment would lead to coagulation and loss of taste properties.

Similar substitutes are being developed, and we believe that the true development of 'light' foods is through the use of these new ingredients, prepared from natural raw materials, whose taste properties have been created so that they best resemble those of high calorie fats.

The implementation of simple chemical reactions in the food industry remains mainly limited to the fats industry; it is digested towards the hydrogenation and the synthesis of emulsifiers and fat substitutes such as Olestra® (sucrose polyester, a non-absorbable synthetic fat product), EPG (esterified propoxylated glycerols), DDM (fatty alcohol ester of malonic and alkylmalonic acids), and TATCA (tricarballic acid esterified with fatty alcohols) [9].

These latter fat substitutes are also developed for the substitution of fats in low calorie foods. They are derivatives resistant to hydrolysis by the digestive enzymes, thermostable and consequently used in bread products, pastries and fried products.

Some of these products have been substituted to the FDA for a commercial licence, others are still undergoing toxicological tests.

It was at the beginning of the century that Normann effected the first hydrogenation with the aid of a catalytic process based on nickel, a procedure which is still in practice today. The study of catalysts and kinetics has enabled this reaction to be controlled and to be steered according to the physico-chemical properties of the fats required. The modification of rheological properties has been made more flexible by the introduction of transesterification and fractionation. Today, the fats industry has at its disposal a wide range of techniques, which enables it to conceive the exact mixture of fats which corresponds best to the desires and specifications of the consumer.

The consumer's attraction to biological processes favours the development and use of yeasts or starter cultures for the preparation of fermented foods. Recourse to known and well-studied strains enables the industry to produce a food which is constant in quality and rich in characteristic flavours.

The preparation of starter cultures, dried by means of different appropriate drying techniques, has given the fermentation industry much more flexibility, and the dairy and cheese industries and the meat industry the ability to better define the aromatic profiles of their products. Genetic engineering will have profitable side-effects on the production of starters, which will consist in modifying the profile of metabolites which condition the organoleptic quality of the product.

Then again, the production of metabolites by microorganisms in intensive culture constitutes another area of research whose results could have interesting repercussions on the use of additives.

Cellular cultures are capable of producing substances destined as food. Take aspartame as an example, a compound which is, at present, playing an important role in the market for sweeteners. Additives of the polysaccharide type can be obtained by exocellular synthesis; these may be composed of only one type of sugar molecule or associations of different molecules. These polysaccharides are used industrially for their viscosity and their capacity to retain water (dextrane, xanthane).

In the fats sector, biotechnology is applying itself to biomodelling particularly of lipids and triglycerides.

The enzyme industry can modify certain enzymes by chemical means so as to modulate certain properties and to adapt these enzymes in order to increase their possible uses. Other possibilities are opened up by progress in genetic engineering in the cloning of genes in microorganisms.

One of the most notable recent successes is the production of recombining chymosine, permitting the replacement and also curing the shortage of natural rennin and coagulating agents.

The food industry constitutes a privileged domain for the industrial exploitation of microorganisms. The use of starter cultures and enzymes enables the implementation of new techniques for the transformation and preservation of food products and consequently contributes to the increase of their added value or intrinsic quality.

## CONCLUSIONS

The obstacle to be surmounted in order to promote the development of new food products is the suspicion of the consumers. Often they will not admit that the food industry is acting in their interest. This is mainly due to their ignorance of the processes and techniques which are industrially applied for the elaboration of commercialized food products. Better relations are necessary; it would be advisable to inform the public at large by trying to explain what is manufactured, how we go about it and the reason for the use of different additives. It is awareness which will create conditions of mutual understanding and that is in the interest of both parties.

Each European nation has developed its own food legislation in order to ensure the production of healthy food. Unfortunately, harmonization of European directives encounters numerous difficulties. The different sectors of the food industry adapt their national and international manufacturing to these directives and free trade cannot be guaranteed until one sole European legislation has come into force. In this context, the food industry has to take into account the wishes of each individual nation's consumers, whose habits are well anchored and difficult to change.

Each country or region will keep its know-how in food, and it will be a challenge to the food industry to producce both quality foods for large international consumption and quality foods which answer to the regional criteria of consumers.

## REFERENCES

[1] Skudder, P. J. (1988) Development of the ohmic heating process for continuous sterilization of particulate food products. Proceedings *Progress in food preservation processes*, volume 1, International Symposium, CERIA, 12–14 April, Brussels.

[2] Ladwig, H. & Depta, N. (1990) Ohmic heating — a continuous sterilization method for media with particulate products. *ZFL* **41** 166–170.

[3] Hermans, W. F. (1988) 'In-flow' fraction specific thermal processing (FSTP) of liquid foods containing particulates. Proceedings *Progress in food preservation processes*, volume 1, International Symposium, CERIA, 12–14 April, Brussels.

[4] Hermans, W. F. (1989) 'Single flow FSTP' — Fractie-specifieke warmtebehandeling volgens het doorstroomprincipe van vloeibare voedingsmiddelen met vaste delen met behulp van het STORK STERIPART systeem. *Voedingsmiddelentechnologie* **31** maaart no. 7.

[5] Müller, H., Moya, J. & List, D. (1990) Direct-sterilization/pasteurization — An optimized HTST–Aseptic process for liquid food containing particulates. *ZFL* **41** EFS 13–19.

[6] Messer, R. (1981) Die Erhitzung von Lebensmitteln mittels direkter Dampfinjektion. *ZFL* **32** 1–4.

[7] Diehl, J. F. (1988) Food irradiation and combination processes: the state of the art. Proceedings *Progress in food preservation processes*, volume 1, International Symposium, CERIA, 12–14 April, Brussels.

[8] Couvercelle Halbwachs, C. (1989) L'ionisation des denrées alimentaires. *Ann. Fals. Exp. Chim.* **876** 159–168.

[9] Summerkamp, B. & Hesser, M. (1990) Fat substitute update. *Food Technology* **44** March 92.

[10] Akahane, T. & Cheftel, J. C. (1989) Surimi et analogues des produits de la mer. *Ind. Agr. et Alim.* **106** 881–897.

[11] Gould, G. W. (1989) *Mechanisms of action of food preservation procedures*, Elsevier Applied Science, London and New York.

# III.4

# Impact of new food concepts on food regulation

C. Crémer

Ministère de la Santé Publique et de l'Environement, Inspection des Denrées Alimentaires, Brussels, Belgium

## INTRODUCTION

The 'new food' concept is an idea that has been emerging in Europe over the past few years, particularly from the moment that biotechnology became available for food development and new ingredients were discovered, such as those intended to replace fats.

However, the development of new foods is not a recent phenomenon. Remember, for example, the appearance of unicellular proteins, vegetable proteins and modified starches, and the introduction of use of food irradiation.

Going further back in time, we find products such as margarine and processes such as deep freezing and even appertization.

In the past, the new ingredients and the novel methods of processing foods were often used empirically, and the various national rules and regulations were established on a case-by-case basis, often in a chaotic manner.

On the eve of the single European market, it is imperative for the EC to have up-to-date and efficient regulations available to protect consumers effectively and to enable the food industry to develop harmoniously.

In the course of this chapter, I intend to bring forward some ideas on the way in which such regulations could be conceived.

## WHAT IS MEANT BY 'NEW FOODS'?

The new foods can be divided into six categories:

— new foods to be consumed as a replacement for the traditional foods; they are obtained from known ingredients and are processed by traditional manufacturing methods. Classic examples are the so-called 'light' products;

— new foods derived from the plant world that can be consumed either as such or after treatment with traditional manufacturing processes. An example of these is the algae, the commercilization of which in some countries has gradually been growing in importance. Over the last few years, shops in Europe have been offering an increasingly wide assortment of hitherto unheard of fruits and vegetables;

— new ingredients, generally produced from plants or other foodstuffs whether or not produced by traditional manufacturing processes. Polysaccharides such as inulin or compounds like that fat substitutes 'Simplesse' or 'Olestra' are typical examples;

— new foods manufactured through novel manufacturing and processing methods. Fruits and vegetables grown through biotechnological means belong to this category;

— traditional food ingredients prepared by using new manufacturing processes such as biotechnology;

— new additives, whatever their origin or method of manufacturing.

## WHAT REGULATIONS SHOULD GOVERN THESE NEW FOODS?

Setting up a policy covering all those new foods if all previous regulations could have been ignored would have been an arduous enough task. But the problem has been made more complex by the existing Directives of the European Community, which cannot be ignored and are very difficult to amend basically.

A coherent food policy should take into account the basic rules underlying consumer protection and information, while acting as an incentive to research and development and guaranteeing fair competition.

These days, such a policy cannot be based on a nitpicking and systematic state control but rather on the principle of the authorities monitoring the self-control applied by the economic operators.

For the six classes of new foods mentioned above, I suggest three different approaches, depending on the case.

The first approach relates to the foodstuffs of the category of which, for example, the light foods belong.

The second focusses on all the other categories except the additives.

The third approach covers the new additives.

## FIRST APPROACH

The new food products to be consumed instead of the traditional foodstuffs, obtained from known ingredients through classic manufacturing processes, should be included in the European regulations that are currently in force or in preparation; in the same way as the traditional foods, they should not be subjected to prior notification or authorization, except in the case of a food intended for a particular diet, for which such a prior notification or authorization is required under Council

Directive 89/398/EEC of 3 May 1989 on the approximation of the laws of the Member States relating to foodstuffs intended for particular nutritional uses.

I am convinced, in the context of the accomplishment of a single European market and the corresponding deregulation, and in line with the Benelux authorities, that a foodstuff should be subject to a specific vertical Directive laying down rules for its composition only in the case of a food that is nutritionally important for consumers or for traditional foods which consumers expect — perhaps instinctively — to have a well-defined composition. In this view, the new foods to which this approach applies should not be subject to specific vertical regulations governing their composition except when it turns out that they are not adapted to the nutritional needs of consumers.

Their labelling, by contrast, should meet all the relevant rules, should include the percentages of basic ingredients and avoid being misleading.

Of course, all the existing Directives and those that may relate to them will also be applicable (e.g. nutritional labelling, hygiene, contaminants, etc.).

## SECOND APPROACH

Foodstuffs and food ingredients which must really be considered as novel because they are foodstuffs or ingredients that have never been consumed before (or at least in a very small quantity as a normal component of a traditional food), or because they are produced through a new manufacturing process, should be treated differently.

Several Member States of the European Community have recently adopted regulations covering these products or are currently studying their introduction. It is necessary, therefore, that the EEC should enact approximated rules and regulations in this respect.

As a matter of fact, the EC Commission has drafted a proposal of joint provisions. This project covers the novel ingredients and the new manufacturing processes, though not the new foodstuffs consumed as such. Under these provisions, for example, algae consumed as such would not be covered by the draft proposal; neither would a new plant consumed as a fruit or vegetable. The draft proposal defines a new ingredient as follows:

> 'one which in the Community has not been used hitherto for human consumption or which has been consumed in only small amounts or has not been used for that purpose bearing in mind considerations given in an Annex.'

These provisions will not be applicable to flavourings or to additives.

I feel that such approximated EC provisions should apply to new foods as well as to new ingredients. Indeed, the new foodstuffs would confront consumers with the same or even more serious problems — whether they be toxicological or nutritional — than the new ingredients; even more so, because they tend to be consumed in far greater quantities.

The approach that I favour would therefore apply to each new food as well as to each new food ingredient.

This approach ought to be flexible enough to allow a harmonious development of the food sector, because each case is a particular one with its own specific minor or major problems.

While being flexible, it should also be effective in providing consumers with the protection they are entitled to expect.

The Community provisions that I advocate would be incorporated into a Council Regulation based on Aritcle 100A of the Treaty. The basic ideas can be outlined as follows.

### Basic criteria

New foodstuffs and ingredients put on the market must in the first place not be harmful to the health of consumers. Neither should they have nutritional disadvantages for consumers. Their labels must be sufficiently transparent and precise enough not to be misleading.

### Notification

The person responsible for putting a new foodstuff or ingredient on the market for the first time should be obliged to assess any possible risks or to have them assessed.

A dossier of the evaluation should be submitted to the EC Commission. It should be based on solid scientific evidence and should focus on the finished product.

The product should only be allowed on the market for the first time some six months following this notification.

During this period, the Commission could, in close cooperation with representatives of Member States, examine the dossier thoroughly and, if necessary, consult appropriate scientific authorities, such as the Scientific Committee for Food, assisted by experts in the field.

If, on expiry of this term, the Commission has not expressed an unfavourable opinion, the new foodstuff would be allowed on the market of all EC Member States under the existing regulations.

At the end of this period, the Commission should be entitled to require an additional period for carrying out an in-depth investigation or to request some additional information.

Any marketing ban and any restrictions on the use of the foodstuff or of the manufacturing processes should be imposed by the Commission in accordance with the opinion of the Standing Committee on Foodstuffs taking the decision by a qualified majority in accordance with procedure III, variant (a) of the Council Decision of 13 July 1987, establishing the modalities of the competences transferred to the Commission.

The marketing of new foodstuffs and new ingredients and the authorization to use a new manufacturing process could, where appropriate, be subject to harmonized community regulations.

Such provisions would also contain a safeguard clause allowing Member States to exclude from their territory during a given period any new foodstuffs or ingredients that would be commercialized without the relevant notification or that would be marketed while there are serious doubts as to their wholesomeness. During this period, a decision would be taken by the Commission after consultation with the

Standing Committee on Foodstuffs. Within a time period laid down in the regulations, Member States can submit the decision to the Council in accordance with Article 3 of the Council Decision of 13 July 1987 mentioned above.

**Some samples of new foodstuffs or ingredients or of new processes that should be the subject of notification**
The indications are as follows:

— the new foodstuff, irrespective of its origin, has never been part of the diet in Europe;
— the new foodstuff, irrespective of its origin, has never been used for nutritional purposes or has only been used in small quantities in terms of the various uses envisaged;
— the new ingredient has been consumed only in small amounts as a substance that naturally occurs in a traditional foodstuff;
— the new manufacturing process greatly alters the nutritional qualities of the foodstuff or causes its nutritional components to deteriorate;
— the new manufacturing process creates undesirable substances in such large quantities as to pose a risk to the health of consumers;
— the new manufacturing process requires scientific research to ensure that it does not pose a hazard to the health of consumers.

**THIRD APPROACH**

This approach covers the new additives, including the extraction solvents. These substances require a special approach, not only because of their special characteristics but also because they are already covered by a number of specific provisions at EC level. The latter includes Council Directive 89/107/EEC of 21 December 1988, on the approximation of the laws of the Member States concerning food additives authorized for use in foodstuffs intended for human consumption, and Council Directive 88/344/EEC of 13 June 1988, on the approximation of the laws of the Member States on extraction solvents used in the production of foodstuffs and food ingredients.

These regulations are based on the principle of the positive list, i.e. all uses are forbidden unless they are explicitly authorized.

This means that, within the EC, there is a strict legislative approach which, in my opinion, should not for the time being be changed.

The EC project on new ingredients that I have already referred to excludes these substances from its scope, and it also excludes the flavourings because they are dealt with in Council Directive 88/388/EEC of 22 June 1988, on the approximation of the laws of the Member States relating to flavourings for use in foodstuffs and to source materials for their production.

Article 5 of this Directive stipulates that under the procedure laid down in Article 100A of the Treaty, the Council shall draw up appropriate provisions regarding the various types of flavourings. Therefore, it is on this basis that the competent services of the Commission envisage adopting regulations governing new flavourings.

In my opinion, it would be preferable to include the flavourings in the scope of the project on the novel ingredients, as the flavourings sector is technologically highly innovative. These substances should be covered by the same provisions as those that are applicable to the new food ingredients and the new foods.

That is why the flavourings should be covered by the regulations which I described in my second approach. This should not cause any legal problem since I suggest that the regulations governing the products covered by the second approach should precisely be based on Article 100A.


## CONCLUSIONS

The difficulty of dealing with new foodstuffs and ingredients and with new manufacturing processes is not itself new. What seems important and urgent to me is that we abandon our empirical ways of addressing the problem and set up at the EC level a general structure capable of appropriately evaluating each new product and each novel process.

For new foodstuffs, in fact, a new regulatory concept has to be developed, in accordance with which the wholesomeness of the product after the introduction of an innovation at whatever stage of the manufacturing process can be assessed by the manufacturer in a harmonized way, while at the same time the Commission of the EC, in cooperation with the member States, is afforded the possibility to check if this was done.

This presentation which I had the honour to deliver to you was restricted to food regulation, but it is self-evident that there are regulations governing other sectors which have a direct or indirect bearing on the possible production of certain types of foods and on the way in which they are produced, manufactured or packed. Examples of such regulations are those governing agricultural policy and the environment.

Environmental regulations could imply that some particular materials would no longer be available for packaging foods.

The use of bovine somatotropin could be prohibited, not so much to protect the health of the animals or consumers, but rather to contain milk production within the context of the agricultural policy. For the same reason, one could imagine that new foods intended to substitute for dairy products would be banned.

The current EC ban on hormones in animal feedstuffs is not only based on the protection of the health of consumers.

The above examples clearly show that when a manufacturer wishes to produce and market a foodstuff, he is obliged to take into account other factors besides the food regulation alone.

# III.5

# Summary report — Food policy trends in Europe: technological aspects

**P. Tobback**
KU-Leuven, Belgium

In Part III, dealing with the technological aspects of food policy trends in Europe, four topics have been discussed.

(1) Trends in the perception of food quality by the consumer.
(2) Opportunities offered for product development innovations in the field of raw materials and food ingredients.
(3) Application of modern techniques in food production.
(4) Impact of novel-food (new food) concept on food regulation.

What follows is a report and comment on these topics.

## 1. TRENDS IN FOOD-QUALITY PERCEPTION

Ever since the scientific achievements resulting from breakthroughs in fundamental research in physics and chemistry were applied in industrial processes in general and food processing in particular, the consumer has been concerned with the quality of industrially produced foods.

The problems existing today derive to a great extent from the fact that the perception of quality by the consumer is strictly 'individual' and strongly influenced by cultural and socio-economic elements. Food quality therefore carries in its perception objective and subjective elements (Fig. 1).

The quality of a food product has been defined as 'fitness for use'. However, in its conciseness this definition carries a high complexity. As objective elements, nutritive value and hygiene-aspects (as related to food safety) are taken into account, whereas

QUALITY: *Fitness for Use*

\* \* \*

ELEMENTS IN QUALITY EVALUATION

*Objective elements:*
— nutritive value
— hygiene characteristics (food safety)
  • microbial load
  • purity: — toxicants
           — contaminants
  — ...

*Subjective elements:*
— sensory
— social–cultural
— economic

Fig. 1 — Food-quality concepts.

the subjective perception of quality comprises sensory (organoleptic) and convenience elements.

The objective elements of quality can be quantified (e.g. in terms of vitamin content, fatty acid ratios, energy content, microbial load, etc.) and are therefore easily used by experts in the field, though unfortunately not by the average consumer. For his evaluation of quality the latter has to rely on purely subjective elements, for which quantification is difficult, even impossible.

Recently, as a result of the consumer's concern about the environment, the quality of a food product is also envisaged against the background of agricultural production (e.g. use of fertilizers, pesticides and hormones) or production of waste. These elements are taken into account by the consumer to attach a supplementary 'image' of quality to the food.

It was concluded that, as a result of the consumer's concept of food safety, which often has a character of 'absoluteness', and owing to the fact that the consumer is very reluctant to accept the concept of a calculated risk, an increased transparency of the food production chain is necessary, together with well-conceived ways of communicating with the consumer in order to bridge the gap between production, transformation, distribution and ultimately, the consumers.

However, the dialogue between scientists or industry and the consumer remains very difficult, and despite the many attempts made and ways used so far, transmitting an honest message to the consumer, when it comes to food, often results in failure. The consumer's attitude towards the use of food additives and his attitude towards E-numbers despite many information campaigns can be cited here as examples.

It was further agreed upon that in the field of food and nutrition the way to communicate with the consumer should be through education. However, the answers to the questions of who should do this and how it should be done were left open for reflection. It is the author's belief that concerted actions on a Community level, like the one started in the FLAIR-programme, should, in respect of this matter, be considered for the future.

## 2. OPPORTUNITIES OFFERED FOR PRODUCT DEVELOPMENT BY RAW MATERIALS, FOOD INGREDIENTS, NEW TECHNOLOGIES

In their origin, industrial production processes of foods were an industrial extrapolation, on an empirical basis, of culinary habits or established craftsman practices. However, with the development of food chemistry and chemical-analytical techniques, data on and insight into the relationship between raw materials composition, processing conditions and end-product quality emerged.

Parallel to this process, the ever-widening gap between the area of food production and the area of food cunsumption has led, on the one hand, to the optimization of known, or the development of new, preservation methods, and, on the other hand, to the increasing use of chemical substances to safeguard product quality and extend shelf-life.

Further, recognizing the constraints and the limitations in raw material composition, technologies have been developed for the isolation of ingredients from these raw materials and the subsequent recombination of these purified ingredients to produce new foods that meet better the objectives of product innovation. This has allowed industry to create foodstuffs to meet specific purposes.

The food production-line as it operates today can be schematized as shown in Fig. 2. Starting from raw materials, foodstuffs are produced about which the consumer

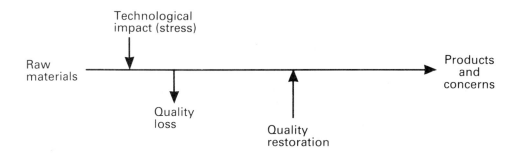

Fig. 2 — Food production today.

sometimes has his doubts. To produce these foods an inevitable technological stress has to be put on the raw material, resulting most of the time in a loss of quality. This quality loss can however be limited, or at least partly restored, by optimizing technological processes or by the use of chemicals (additives).

Further, it can be stressed that the enormous variety of food products now presented to the consumer in the developed countries has led to a new trend related to the requirement to strike a balance between need and pleasure. Food ingredient substitutes for basic elements like sugar, fat and proteins have been developed and incorporated into existing food-products to reduce, for example, their energetic value.

The potentials of new methods in food preservation to reduce losses in quality (such methods as ohmic heating, microwave heating, fractional-specific thermal-processing (FSTP), high-pressure treatment and microparticulation) together with the potential use of genetic engineering and bioprocessing have been recognized.

The food production-line as it will, or may, operate tomorrow is schematized in Fig. 3. Very likely the food industry will have at its disposal raw materials, the characteristics of which have been optimized by modern genetic techniques and in which particular quality attributes are enhanced. Using modern bioprocessing techniques more adequate foodstuffs will be produced in which undesired factors are eliminated. However, even optimized bioprocessing will put a stress on the raw materials and some quality loss is to be expected.

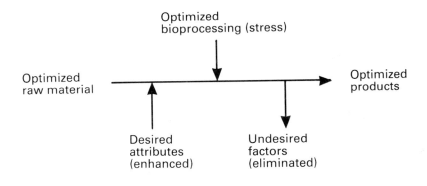

Fig. 3 — Food production tomorrow.

The potential consequences of the above-cited technologies can be very wide. They may lead, for example, to a reduction in the use of certain additives in foods, a reduction of the presence of natural toxicants or to an enhancement of the nutritional value of the foodstuffs.

## 3.   NOVEL-FOOD CONCEPT AND FOOD REGULATION

The 'novel-food' concept has emerged in Europe during the past few years, more specifically from the moment that modern biotechnological processes became available in certain areas of food production. The concept is increasingly used, but its content is still ill-defined.

It should be emphasized that the development of what one might call 'new foods' is a phenomenon that is not particular to the present moment. Evolution in technologies in the past has always led to the development of 'new' products. However, as already indicated above, biotechnology adds a new dimension to the concept of 'novelty'.

Six categories of novel foods can be distinguished:

(1) Replacement foods for traditional ones, obtained from known ingredients;
(2) New products derived from the plant world and, up to now, not used as foods;
(3) New ingredients from plant or other foodstuffs;
(4) New foods produced using new methods;
(5) Traditional ingredients produced by biotechnological processes;
(6) New food additives, independent of their origin.

As regards legislation, essentially two approaches can be considered: (i) where the so-called novel foods are replacements for traditional foods but are still obtained starting from known ingredients (category 1), and in the case of new food additives (category 6), regulation can be performed according to the existing mechanisms or principles; (ii) for food developments for which there might exist some basis for concern as to potential risks (categories 2–5) the following principles should be considered:

(a) Novelty should not in itself be considered as a risk factor.
(b) During the process of development of the food, the principle of 'autocontrol' should be put into practice. By autocontrol is meant that together (i.e., at the same pace) with the development of the 'novel food', safety assessment should be carried out. This means that at the end of the development process, the data on safety evaluation of the product should be readily available.
(c) Notification should be given to the appropriate regulatory body. This notification is required in order to enable the regulatory body to check (i) if a safety assessment was carried out, (ii) if this assessment was carried out by qualified authorities.

It was further emphasized that regulations governing other sectors (e.g., regulations with regard to Community agricultural policy or in relation to environmental problems or issues) may have a direct or indirect impact on the production of novel foods or on their manufacturing or packaging methods.

# Part IV
# *Food control and analysis*

# IV.1

## Fast analytical methods in chemical food analysis

**Werner Baltes**
Institut für Lebensmittelchemie der TU Berlin

The centennial of Belgium Food Regulation may be a reason for some thoughts dealing with the development of food regulations and analytical possibilities since that time. The first food laws issued in that time might mostly have been dealing with prohibitions of poisonous ingredients in food. In this sense they might have followed the tradition of older local regulations to keep citizens protected from spoiled and poisonous foodstuffs. For example, laws were published in Germany concerning the use of lead- and zinc-containing colours and articles for consumption in 1885 and 1887. But the new food laws also contained regulations about the composition of food in order to protect the public from fraudulent practices.

These regulations required analytical methods able to check the food. How far had development progressed by 1890? It is well known that the nineteenth century is characterized by numerous discoveries in the field of natural sciences. For example, it had begun to be recognized that organic chemistry could be ruled by man, whereas in earlier times people believed that such compounds could be formed only by nature. Now a development of synthetic and, last but not least, analytical methods in this field occurred. In the middle of the century, J. v. Liebig had founded his agricultural chemistry. In 1890, Emil Fischer in Berlin solved the constitution of sugars and Louis Pasteur was the grand director of his famous research institute in Paris. Baking powder was not yet discovered, but already milk powder (Nestlé 1860) as well as skim milk were available on the market. (The first factory for the production of baking powder was erected in Switzerland in 1866.) In 1810 a British admiral developed a special oven for the automatic production of crackers.

Josef König, first a professor of agriculture in Münster/FRG, began subsequently to analyse food and to collect the knowledge in this field which had already been published by others in the nineteenth century. In 1878 he first published his comprehensive book about the chemistry of man's nutritive and semi-luxury foods (*Chemie der menschlichen Nahrungs- und Genussmittel*). The tables of this book

(3rd edition 1903) contain values for the water, nitrogen, fat, fibre and ash content of a great many foodstuffs. Wine analysis contained tests for specific weight, alcohol, extract, whole and volatile acid, tartaric acid, sugar, glycerin, ash and phosphorous acid. When there was no method for the determination of a particular compound, a compromise was elaborated to judge the food. One of the most difficult problems was the analysis of fat, in spite of the fact that very intensive research had been done in that field in that century. It was well known that fats are mixtures of different glycerides. The analysis of them proved to be very complex. So our predecessors elaborated standard methods for a summary fat analysis in the first years of this century, some of which are still in use:

| | |
|---|---|
| free fatty acid number | (in German: *Säurezahl*) |
| saponification number | (in German: *Verseifungszahl*) |
| iodine number | (in German: *Jodzahl*) |
| melting point | (in German: *Schmelzpunkt*) |
| solidification point | (in German: *Erstarrungspunkt*) |

Polenske, who elaborated some other standard methods in the following years, distinguished between tallow and lard by special differences between their melting and solidification points. In the exercise of these pragmatical methods which were used as a preventive consumer protection, the users (analysts) became very specialized experts. So, in 1894 in Germany, 'official chemist' was established as a special occupational group. This arrangement has functioned very well.

It is well known that food chemistry passed through a boom until 1930. During this time a great many procedures for food analysis were elaborated. A second boom resulted from the enormous development of physical methods in chemical analysis in the years after 1960. Today, most of our foodstuffs can be analysed extremely thoroughly, and the detection limits have been lowered to the region of ppb, ppt and ppq. By the use of modern methods of food analysis most actual problems can be solved. At the same time, a lot of the methods are laborious and time-consuming.

The judgement of food requires a considerable number of analytical parameters, which makes a high demand on effort. For example, knowledge simply of the composition of a food is not sufficient for us: we also have to check the origin of its ingredients, any food additives, production aids, contaminants and agricultural residues. So, different types of analytical network have to be applied.

When a foodstuff of unknown recipe has to be tested methods disclosing the basic composition are applied first. In this way, the percentage of water, mineral, fat, protein and the sum of carbohydrates are determined. Subsequently, special analytical methods are carried out to determine the percentage of special ingredients like alcohol and vitamins, the amount and type of carbohydrates, the fat byproducts and other constituents. Finally, the kind and concentration of food additives and residues from the environment or from food production have to be determined. The more detail we see, the smaller the amounts become and the greater the effort needed for their analysis. But each of these compounds has to be determined exactly, and if one of them is overlooked a new scandal can possibly occur.

Thus it is clear that fast methods are required, which give an overview of some parameters of the food very rapidly.

What is the definition of fast methods? Such procedures may be defined according to:

(1)  *The equipment*
(a)  low-cost procedures (test-kits, DC, potentiometric methods)
(b)  instrumental methods with possibly high costs (HPLC, NMR or NIR)

(2)  *The kind of method*
(a)  detection and screening methods
(b)  tests or limiting value methods
(c)  quantitative or semi-quantitative methods
(d)  methods of process control.

As the last-named is independent of the expenditure, it seems to be the most useful. In this connection, Matissek [1] has published a graphic, which is shown in Fig. 1.

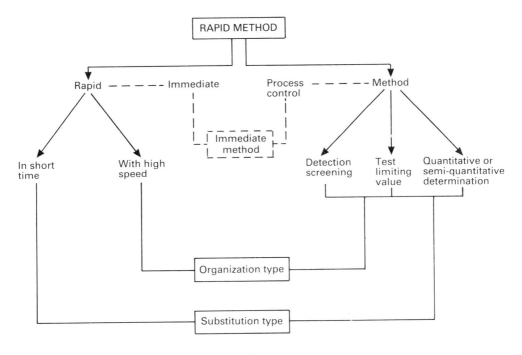

Fig. 1

The substitution type of method is characterized by a change in procedure in favour of faster working methods. For example, the method of extraction and gravimetric determination of the fat content can be substituted for by a procedure which includes a low-resolution NMR instrument (~20 MHz) which is able to determine the fat content in oil fruits without any extraction. Another type of rapid method is represented by a principle which works faster than the traditional principle. An example would be the application of the 'Kjel-Foss Automatic' instead of the Kjeldahl method. Another example is the use of a Technicon instrument performing tests in large numbers automatically, while the basic principle is retained. Last but not least, immediate methods can be used for process control. An example is the measurement of protein content in milk products by NIR (Near Infrared Spectroscopy), which can be made quickly parallel to the production process by coupling the IR-instrument with a computer. Other applications of NIR have been described which enable measurement of moisture content in cereal flour [2], of sugar content in fruits [3], of theaflavin [4] and fibre [5] in black tea, and of moisture, oil, protein and glucosinolates in rapeseed [6]. In every case, these on-line measurements yielded values of sufficient accuracy without time-consuming pre-preparation of the food. This method is based upon measurements of molecular vibrations (overtones or combinations) in the region 750–2500 nm (corresponding to 400–13 000 cm$^{-1}$). The most intensive bands correspond to CH-, NH- or OH-stretching vibrations, the last-named of which yield relatively wide and low absorptions. Therefore the spectra of food in this area possess the character simply of fingerprints, which is less valuable for interpretation of a structure. This means that NIR needs a semi-empirical calibration which has to be controlled statistically by a computer. On the other hand, the low absorption causes a relatively great penetration, so that average values for a bigger volume can be yielded. Because of the robustness of the instruments, the use of NIR methods (mostly applied in the reflectance version) is one of the most interesting developments of the last 20 years [7,8].

Other principles of in-line, or on-line, controls of food production are based on sensors yielding values such as temperature, weight, pressure, flow and pH. Especially pH measurements of waste water can be very important because of environmental protection. Further methods for control of food production are HPLC, GC and FIA (Flow Injection Analysis) where sampling can be accomplished by placing the sensor in a bypass of the material stream. The more precisely these instruments function the higher is the safety in quality control during food production, saving analytical work in the laboratory.

A very special development in fast food control is the the image analysis, which can be placed in-line in the food production process, without direct contact with the food. The basic idea of this method is the radiation of food by visible, UV, IR or X-ray radiation and simultaneous registration of the resulting image by a video camera coupled with a computer where the signal is digitized and stored [9]. One example of its application is in the fish industry, in the production of boneless fillets. This methodology takes advantage of the strong autofluorescence of fish bones at 390 nm after radiation with UV light of 340 nm [10,11]. The method is applicable to various fish species (cod, haddock, plaice, salmon and whiting). An instrument has been constructed for classifying fish fillets and seems to be available [12]. It works with a speed of 1 fillet/second.

The same principle is applied to measurements of the fat/lean ratio and of the amount of connective tissue in meat products. Chemical analysis of meat products is rather laborious. An in-line control of pork, beef and poultry can be carried out very easily by radiation with UV light of 340 nm and recording the fluorescence at 390 and 475 nm by image analysis. Cartilage, too, can be detected by a fluorescence of 390 nm; this seems due to elastin fibres [13,14]. Another method of fat determination in meat is based on X-ray absorption, which is a function of mineral content. Because the mineral content in fat meat is lower than in lean meat, this method is suitable for measurement of big pieces of meat (7 kg sample) as grown [15,16].

There are available a great many other measurement principles which could be used by the food industries (e.g. electric and biochemical sensors). It is a challenge for the food industry to encourage the construction of suitable instruments for the control of food manufacture [17].

Rapid methods are also needed in the laboratory. The number of samples rapidly increased so enormously that the development of faster working methods of equal accuracy became necessary. Many of them were derived from the traditional procedures, other methods utilized new principles.

As an example, the methods for fat determination in food, shown in Table 1, will be explained. Fat is determined basically after its disintegration in the sample by means of extraction with suitable solvents. In the classical methods, the solvent is evaporated and the residue is determined gravimetrically after drying. A substantial acceleration is achieved by measuring the density of the fat solution in a suitable manner. The Foss-kit method realizes this step by means of a buoyant body which

**Table 1** — Substitution of traditional methods

| Parameter | Traditional method | Rapid method |
|---|---|---|
| Fat | Gravimetry after solvent extraction<br>Gerber method | Densitometry<br>Refractometry<br>Volumetry<br>Turbidimetry |
| Protein | Kjeldahl | Kjel-Foss<br>Direct distillation<br>Colour binding<br>Thermal conductivity detector |
| Solid matter/moisture | Gravimetry after thermal drying | IR-drying<br>Microwave drying<br>Conductometry |

basically corresponds to an areometer. In order to increase accuracy and to facilitate the reading, these instruments are equipped with a digital display which may also allow the printing out of the values. One test needs 5–10 minutes [18]; the sample (weight 45 g) is homogenized in a special cell and then mixed with 120 cm$^3$ of solvent and some gypsum for binding the moisture.

Another principle of fat determination is based on refractometry. This method is especially used with α-bromonaphthaline, the refraction index of which is influenced very sensitively by the concentration of dissolved fat. This method is used for fat determination in cocoa products, confectionery and meat [19]. For large numbers of liquid samples, volumetry works very well. The fat is disintegrated by sulphuric acid and driven by centrifugation onto a layer of amylic alcohol, the volume of which is increased by the fat amount. This determination is carried out in vessels of special size allowing an accurate reading of the volume increase. This method, which was developed by Gerber in 1892, is still in use [20]. It offers a fast method for milk analysis, because a large number of samples can be analysed simultaneously. To carry out one run, about half an hour is necessary. A new method represented by the Milko-Tester provides for the elimination of milk turbidity caused by casein by means of 'versene solution'. The turbidity of fat in the water phase remains, and this is determined. This method takes 45 seconds per sample [21]. In the field of protein analysis also, some fast methods are known. For example, a modification of the traditional Kjeldahl method resulted in the development of the so-called 'Kjel-Foss-Automatic' method. The single steps of Kjeldahl analysis (decomposition of the sample by $H_2SO_4$, distillation of free ammonia from an alkaline medium) are arranged on a carousel which is able to start a new analysis every three minutes. The endpoint of the titration is measured photometrically. The time taken for the test is 15 minutes, and 20 samples can be analysed in 1 hour [21].

The method of direct distillation is based on the fact that the reaction of strong alkali on protein causes in the first place the elimination of ammonia from the acid amide groups of glutamine and asparagine. This ammonia is distilled and determined as in the Kjeldahl method. The method of direct distillation is limited to seeds, cereal and meat products. It requires an extensive number of standards analysed by the Kjeldahl method to make the correlation to the total nitrogen content of the sample reliable [22].

Colour-binding methods are well suited to milk testing. The basis of this so-called 'Amido Black method' is the adsorption of certain colours on protein. Under standardized conditions the amount of dyestuff not adsorbed can correlate to the concentration of protein in the sample [23].

The method of thermal conductivity measurements is based on the conditions of organic elementary analysis, which also have been modified. Generally, the sample is burnt in conditions under which it is converted to $CO_2$, $H_2O$ and $N_2$. The modifications, on the market since about 1965, guarantee accurate values because of the total decomposition of the sample at 1000°C. The decomposition products are separated, using helium as the carrier gas, and are measured by means of a thermal conductivity detector. This principle is also applied to protein analysis. The test period is about 10 minutes per sample [21].

The methods cited for solid matter and moisture determination need no detailed explanation. It is clear that the test periods can be shortened by the application of IR-drying or microwave drying. Also the electrical conductivity of a sample depends on its water content. Of course a sufficient number of standards, measured by other methods, must be available [21].

First priority during a discussion of modern principles for food analysis must be given to separation methods. Chromatography made possible for the first time the direct identification and quantification of numerous food ingredients which previously had been determined indirectly or by means of analytical standard 'numbers' as in the analysis of fats. Most important for food chemistry are TLC (thin-layer chromatography), GC (gas chromatography) and HPLC (high-performance liquid chromatography) with their variants. Among electrophoretic methods capillary isotachophoresis has to be mentioned. All of these methods are suitable as fast methods for screening as well as for quantification (see Table 2). These methods are really the province of the modern food chemist, who knows their principles. Therefore only some special remarks seem to be necessary.

**Table 2** — Separation techniques

|  | Thin-layer chromatography | Gas chromatography | High-performance liquid chromatography | Isotachophoresis |
|---|---|---|---|---|
| *Screening*: | By $R_F$-value and colour of the spot | By $K_I$-value (2 different columns recommended) or by coupling of GC with MS | Via elution time and diode array detector | Possible with limited choice of analytes |
| *Quantification*: | TLC-scanning (remission spectrophotometry) | By integration |  | By photometric or other detectors |

Each of the methods mentioned is very sensitive to by-products and impurities. Therefore pre-cleaning is mostly necessary. TLC can be applied for identification as well as for quantitative determination of compounds. In the latter case the laboratory must be equipped with special scanners working on the principle of reflection spectrophotometry (remission spectrophotometry). Quantitative TLC requires reproducible plates concerning the thickness of their layers, as well as very accurate and reproducible application of the sample solution. The accuracy of results is comparable with spectrophotometric measurements. Every compound which possesses a specific absorption maximum can quickly be determined via quantitative TLC. Because of a description and publication list see [24].

A special application of GC as a fast method is represented by headspace GC. This method does not require any precleaning because only volatile compounds are applied. On the other hand, this method requires, for quantitative analysis, a sufficient number of standards [25].

HPLC can also be used as a screening method [26]. For the identification of unknown compounds the application of a diode array detector can be recommended. Among rapid methods for food analysis, capillary isotachophoresis has proved very useful because a time-consuming clean-up is not necessary and, in particular, polar compounds, which are not suitable for GC, can easily be separated and determined.

In recent years several methods have been published (for example for the determination of inorganic anions from mineral water, for organic acids, vitamins, biogenic amines and artificial sweeteners). For a detailed description, see [27].

Low-resolution NMR spectrometers have been on the market since about 1965. These instruments, which belong to the class of continuous-wave (CW) NMR attracted the attention of the oil chemists because of their capability to quantify moisture as well as fat content in oil seeds in a matter of minutes.

Now, pulsed NMR instruments are available, which can quantify fat moisture within a few seconds. This technique provides an intense pulse of radio frequency for some microseconds, after which the nuclei return to their original state and send out the NMR signal simultaneously. A maximum intensity is achieved when the nuclei have been rotated 90 degrees with respect to the static magnetic field ($\rightarrow$'90 degree pulse'). Because of different relaxation times, signals due to hydrogen nuclei in solid phases decay more rapidly than those in liquid. Therefore this method is also capable of measuring the portion of liquid oil as well as solid fat (SFI-index). Other applications are the determination of cocoa butter in chocolate powder, moisture and oil in oil seeds, milk fat in skim milk and dried milk, and moisture in starch products, in fat and in confectionery products. There are only few descriptions of this application of NMR, which is better known as a method for structure elucidation (high-resolution NMR). You will find a very useful description in [28].

Fast methods for food analysis can also be based on ion-sensitive electrodes. Basically they are derived from the glass-electrode, which means that the most simple ion-sensitive electrode is represented by the $H^+$-electrode. In ion-sensitive electrodes, different membranes are used which mostly consist of a low-solubility precipitate of the element in question (solid-state membranes): for example, AgCl, AgBr, $LaF_3$ for the determination of $Cl^-$, $Br^-$ or $F^-$ ions.

Liquid-membrane electrodes consist of a non-water-miscible organic solvent on porous plastic discs in which the ionophores are dissolved. Ionophores are complex-forming active compounds such as valinomycin (for $K^+$-electrodes) or crown ethers (for $Ca^{2+}$, $Li^+$ and $Ba^{2+}$). Immobilized enzymes can also be used and form in this way enzyme electrodes for the determination of, for example, glucose (by glucose-oxidase), urea (urease), ethanol (alcoholoxidase). There also exists, for example, a specific electrode for nitrate. Of course, ion-sensitive electrodes, too, require a reference electrode to be active. An introduction to this field is published in [29,30].

Ion-selective electrodes are suitable as detecting units in the flow injection analysis technique (FIA) [31]. The sample is injected via a microlitre syringe into a carrier electrolyte solution continuously transported to the detector. If analyte ions are present a signal peak is obtained.

This survey would be incomplete if test-kits were not treated. The most suitable seem to be test-strips consisting of a reactive test-zone immobilized on a plastic backing. The test-zones are impregnated with reagents, buffers and other compounds. Test-strips are used for the rapid exploratory testing of special compounds and their semi-quantitative determination at concentrations in the ppm range. For example, the nitrate test-strip is graded as 0–10–25–50–100–250–500 ppm. It also contains a nitrite warning zone. Any nitrite present must be destroyed by means of amidosulphonic acid prior to the nitrate determination. One requirement is essential: the compounds being tested for must be soluble in water. Publications about the use of test kits in food analysis deal with Fe, Co, Cu, Zn, Al, As, Ca, Sn, nitrite, nitrate and peroxide. There are also kits for the analysis of peroxides in deep frying fat which allow users of frying equipment to test the quality of their fat.

Looking towards the future, it seems likely that the development of rapid methods for food analysis will increase. With perhaps some exceptions, most methods concerning measurement accuracy, will be in line with the classical standard methods of the official food authorities.

# REFERENCES

[1] Matissek, R. (1990) In: *Rapid Methods for Analysis of Food and Food Raw Material* (ed.: W. Baltes), Behr's Verlag, Hamburg, p. 23.

[2] Williams, P. C., Thompson, B. N., Wetzel, D., McLay, G. W. & Loewen, D. (1981) *Cereal Foods World* **26** 234.

[3] Davenel, A., Crochon, M., Sevila, F., Pourcin, J., Verlaque, P., Bertrand, D., & Robert, P. (1987) In: *Rapid Analysis in Food Processing and Food Control* (eds: W. Baltes, P. Baardseth, R. Norang, & K. Søland), Norwegian Food Research Institute PO Box 50, N-1432 Ås-NLH Norway, p. 171.

[4] Hall, M. N., Robertson, A., & Scotter, C. (1987) *loc. cit.* **[3]**, p. 176.

[5] Yan S., Meurens, M. & Vanbelle, M. (1987) *loc. cit.* [3], p. 186.

[6] Ribaillier, D., Chesneau, L. & Bioteau, P. (1987) *loc. cit.* [3], p. 181.

[7] Davies, A. M. C. & McClure, W. F. (1985), *Anal. Proc.* **22** 321.

[8] Cooper, P. C. (1983) *Cereal Foods World* **28** 241.

[9] Braggins, D. (1986) *Danish Optical Society*, DOPS-NYT, **4** 8.

[10] Jensen, S. A., Munck, L., Sigsgaard, P. S. & Huss, H. H., *American patent nos* **4**, 631,413.

[11] Huss, H. H., Sigsgaard, P. & Jensen, A. A. (1985) *J. Food Protection* **48** 393.

[12] Jensen, S. A., Lumetech Ltd. Strandvejen 50, 2900 Hellerup, Denmark.

[13] Newman, P. B. (1984) *Meat Science* **10** 87.

[14] Newman, P. B. (1984) *Meat Science* **10** 161.

[15] Seffelaar & Looyen, *Odenzaal Holland (1978)*, Selo Bulletin Ablage 125-5-71-1.

[16] Madigan, J. J. (1978) Information of Anyl-Ray Corp., Waltham MA, USA.

[17] Jensen, S. A. (1987) *loc. cit.* [3], p. 23.

[18] Montag, A. (1973) *Deutsche Lebensmittelrundschau* **69** 470.

[19] Fincke, A. (1965) *Handbuch der Kakaoerzeugnisse*, Springer-Verlag Berlin, p. 428.

[20] DIN 10310 (1970): Bestimmung des Fettgehaltes von Milch nach dem Gerber-Verfahren. *Amtliche Sammlung von Untersuchungsverfahren nach §35 LMBG* (1981) (ed.: Bundesgesundheitsamt Berlin), Beuth-Verlag Berlin, Köln, L 01.00-8.

[21] Torkler, K. H. (1990) *loc. cit.* [1], p. 59.

[22] Schmütz, W. & Do, Q. N. (1979) *Deutsche Lebensmittelrundschau* **75** 398.

[23] Kiermeier, F. & Lechner, E. (1973) *Milch und Milcherzeugnisse*, Verlag Paul Paray Berlin, p. 256.

[24] Jork, H. (1990) *loc. cit* [1], p. 107.

[25] Kolb, B. (1982) *Labor Praxis* **6** 4.

[26] Engelhard, H. (1977) *High performance Liquid Chromatography*, Springer-Verlag Berlin.

[27] Holloway, C. (ed.) (1984) *Analytical and preparative electrophoresis*, Walter de Gruiter & Co, Berlin and New York.

[28] Barker, P. J. (1990) *loc. cit.* [1], p. 267.

[29] Honold, F. & Cammann, K. (1990) *loc. cit.* [1], p. 330.

[30] Mattiasson, B. (1987) *loc. cit.* [3], p. 330.

[31] Ilcheva, L. & Cammann, K. (1985) *Fresenius Z. Anal. Chemie* **322** 323.

# IV.2

# Quality assurance in food analysis: where are we going to?

**Hedwig Beernaert**
Institute of Hygiene and Epidemiology, J. Wytsmanstraat 14, B-1050 Brussels, Belgium

## INTRODUCTION

The quality of a food product depends upon many processes: the art of cultivation of raw materials, the Good Manufacturing systems, the optimalization of transport and distribution facilities, correct labelling and presentation of the product and finally the conditions under which the products are consumed. There is no doubt that texture, smell, taste and price of food products influence to a high degree the consumption pattern of the customer. However, more important is the relationship between the quality of the food market products and their risks for safety and health. To balance these two measurable attributes a quality assurance programme must be available and applied. Therefore, the quality and the quality control of food can only be assured if a management system is created and implemented along well-established international rules.

## QUALITY ASSURANCE STRATEGY IN EUROPE

The realization of the internal market in 1993 involves the free movement of foodstuffs in the Member States without economic barriers [1]. However the protection of public health can justify a complete ban on importing and marketing foodstuffs imported from another Member State where they are lawfully produced and marketed. The presence of undesirable food additives, residues and contaminants, the microbiological deterioration of certain foodstuffs and unclear labelling are public health arguments which can be used by each Member State to forbid the import of foodstuffs. To avoid this problem, harmonization of laws and normalization of systems used in the European Community have to be realized. In the context of the Council Directive 83/189/EEC [2], the European Standards EN 29000 and EN 45000 have been prepared under mandate of the Commission of European Communities and the European Free Trade Association, by a joint CEN/CENELEC Working Group on Certification. The standards are based very much on ISO/IEC guides and on the work of the International Laboratory Accreditation Conference (ILAC). The EN 29000 Standards specify the criteria for the quality assurance of products, services and systems. The criteria set out in the EN 45000 Standards are those to which laboratories should conform and which should be used by accreditation bodies in accrediting laboratories, by public authorities when designating laboratories for regulating purposes and by any other organization assessing laboratories.

In the European Community, the following accreditation bodies are known: Sterlab (The Netherlands), National Measurement and Accreditation System (NAMAS) (Great Britain), Réseau National d'Essais (RNA) (France), The Danish National Testing Board.

In Belgium, a National Council of Accreditation and Certification has been installed. It includes three sections: the Belgian Organization of Calibration, the Belgian Organization for Testing Laboratories and Control Units, and the Belgian Organization for Certification of Products and Systems. As to the accreditation of laboratories for the Department of Public Health, a Royal Decree of 23 October 1965 [3] is in force that sets out the criteria to which laboratories should conform in the field of food analysis. In Belgium, about 40 official, university and private laboratories are certified for the chemical analysis of foodstuffs. To guarantee that these laboratories apply quality assurance, a review of their accreditation status will be carried out in the near future. Objective criteria such as the production of high quality analytical data through the use of analytical measurements that are accurate, reliable and adequate will be involved. During inspection, special attention will be given to the integration of Good Laboratory Practice in the quality system [4].

## INTEGRATION OF QUALITY ASSURANCE IN LABORATORIES

In Belgium, two types of laboratories can be distinguished: the official ones working with the financial support of the Government and the private ones working with their own budget. This situation can influence the cost and eventually the quality of the analysis. Therefore, to assure the quality of food analysis the staff who are responsible for the management of the laboratory must develop the framework for strategic business planning in this environment. The approach has to be built on a five-forces model: marketing, financial impact, investment, exploitation and results (Fig. 1).

Before beginning the installation of a laboratory for food analysis it is clear that a thorough knowledge of the market is required. Special attention has to be paid to the type and the volume of analyses asked and to the competing laboratories. A second important point is the financial aspect. What is the money input I need for the installation of the laboratory? Three procedures can be used: own capital, loan or mixed. A calculation of the return on equity can determine the decision. A third

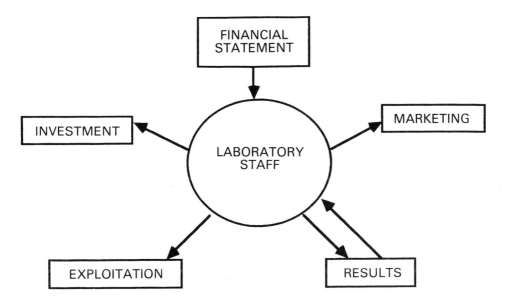

Fig. 1 — Five-forces model.

aspect is the investment. For the smooth functioning of the variety of food analyses which will have to be carried out, the quality assurance programme has to devise an optimal structure of the rooms. A scientific investigation of essential equipment and accessories is a must in order to optimize the investment procedure. The next step is the exploitation of the laboratory. The application of objective criteria such as accuracy, precision, reliability and rapidity of analysis and also the integration of the criteria of EN45000 are the key factors for the implementation of quality assurance in the exploitation of a laboratory. Finally, the quality and existence of the laboratory will be assured if the results are reliable and if the returns on investment and equity are positive at a certain degree.

The concept of quality integration in a laboratory can be illustrated by a graphical model (Fig. 2). When laboratories begin their activities, a certain knowledge is present among the staff (E). With this knowledge a technical maximum of quality integration is possible (D). The most ideal position is point A, where integration and maintenance cover each other and result in a 100% quality system. However, in practice the maximum integration of quality that can be reached is illustrated in point B. Moreover, the laboratories able to follow the bisector EB are exceptional. Nevertheless, a minimum degree of quality integration is required if the laboratory is to be accepted on the market. This will be the case if the laboratory represents a quality curve EF. When a laboratory has a quality curve EG a failure becomes unavoidable.

## PRINCIPLES OF QUALITY ASSURANCE [5]

The introduction of quality assurance programmes in food analysis gives the laboratory the opportunity to produce analytical data of high quality. It means that

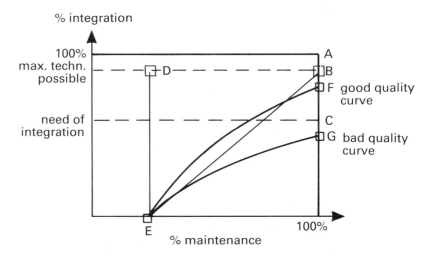

Fig. 2 — Quality integration — maintenance model.

the analytical measurements are accurate, reliable and adequate. To realize this the laboratory needs a management system that controls personnel and time, productivity, the performance and efficiency of the programme and finally the results. The goal will be different from one lab to another and depends on the complexity of operations, the size of the lab, the kind of contracts and the type of personnel present. Therefore, each lab has to develop a plan if the quality assurance programme is to be successful. Generally the plan is composed of three essential elements:

- *prevention* requires a programme planning to ensure that analytical systems are functioning properly (e.g., training, calibration, maintenance, standardization);
- *assessment* is a control component that includes periodic checks on performance to determine precision and accuracy;
- *correction* is an element of action for the purpose of determining quality defects and restoring proper functioning of the analytical system (e.g., trouble shooting, retraining).

When a sample enters a laboratory, the key question to solve is: who has to do what to assure that the quality of the analysis is realized. The analysis is at the heart of the quality assurance programme. Management, training and experience of personnel, techniques, instrumentation, sampling, methodology, intra- and inter-laboratory ring tests and interpretation of data can influence in different ways the conclusions of the results obtained.

In all measurement systems, the analytical laboratory encounters critical control points in the following areas:

- *Sample collection*: Sampling is one of the most common sources of analytical error. The objective of any national sampling project is to determine the state of compliance of selected food commodities in the market place with regard to the selected compounds. In the case of pesticides, an absolute compliance which established maximum residue levels (MRLs) would be assured if a considerable number of samples of every commodity, both domestic and imported, were analysed for every possible pesticide residue. Such a consideration is not practical and realistic. Therefore, the manager has to decide which pesticides should be analysed, which pesticides should be measured in each commodity and how many samples should be analysed for each commodity and pesticide selected. Another important point is the capacity for the analyses in the lab. If we assume that the number of lots of any given commodity on the market is very large, the upper limit of the possible percentage of unsatisfactory lots will vary with the number of specimens analysed. H. B. S. Conacher [6] has calculated that if only 25 samples are analysed and all are satisfactory, approximately 15% of the lots on the market could be unsatisfactory. If, however, 200 samples of each commodity are analysed and all are satisfactory then ca. 2.5% of the lots on the market could be unsatisfactory. To develop such a strategy it is of great importance to know the origin of the products sampled and to have information about the pesticide treatment applied on the different commodities; and, of course, the procedures for collection, preservation and shipment of samples must be taken into consideration.

- *Sampling for analysis*: Once the sample has arrived and the administration task has been done, the analyst has to carry out the type of analysis requested. Therefore he has to decide which analytical portion he needs that is representative of the sample and representative of the present lot. There are two choices: the preparation of a composite sample or the examination of multiple individual units. Sometimes it is desirable to use both types. If homogeneity and variability between units is not of great importance, the procedure of choice will be the analysis of a composite sample. Moreover, this procedure is time-saving. On the other hand, multiple-unit sampling is indicated when the range of individual units is large (aflatoxins in peanuts). A serious mistake is to composite several samples and then to run repeat determinations of this composite sample. Finally, it has been proven that the reliability of the result generally increases with the square root of the number of samples analysed. Therefore, multiple samples are always preferred over single samples, since single samples give no information on the homogeneity of the lot that was sampled.

- *Methods of analysis*: These give the analyst the opportunity to provide reliable information on the nature and composition of materials submitted for analysis. The objective of any quality assurance programme is to hold the variability of the measurements to a minimum. Generally, the variability of inter-laboratory analyses is greater than that of the intra-laboratory tests. The variability of the methods of analysis can be measured by attributes such as accuracy, precision, specificity, sensitivity, detectability, dependability and practicality. All these attributes cannot be measured in optimal conditions at the same time: it takes too much time and it is very expensive. Therefore, for one particular situation the analyst must decide which factors are essential. The method selected should provide the following data:

— evidence of analyte identification
— evidence of separation of the analyte from interfering substances
— lower limit of measurement of analyte concentration
— an accceptable measurement of precision (intra–inter)
— accuracy of measurement of the analyte.

The exactness of the result depends on the management of the laboratory, the analyst, the concentration of the analyte, the type and nature of the interferences or contaminants, the limit of determination and the integrity and stability of the analyte.

In a quality assurance programme several types of methods can be distinguished:

— official methods
— reference methods
— screening methods
— routine methods
— automatic methods
— modified methods.

If there is no legal requirement for using a specific method the analyst has the possibility to choose methods on the following bases:

— matrix effects
— range of concentration of the analytes
— methods available for a lot of food products and a variety of compounds to be identified are chosen over methods with a limited use
— simple, rapid and low-cost methods should be chosen over complex, slower and more costly methods.

The suitability of the methods used has to be tested with the help of reference materials or inter-laboratory ring tests. For the ring tests laboratories well experienced in the field should be chosen.

● *Inadvertent errors*: These errors can happen to both analysts and supervisors. They can and do occur when the GLP-criteria are not rigorously followed.

● *Instrumentation*: The performance of optical and electronic instruments will change as a function of time. Therefore, calibration intervals should be assigned to all equipment as part of the laboratory's preventive maintenance programme.

● *Analysts* have a key function in the operation of the laboratory. Adequate training and appropriate experience are required of the analysts if quality assurance in food analysis is to be achieved. The analyst must also have knowledge of the corrective action plans for when errors, deficiences or out-of-control situations occur.

- *Validation of results* [7]: Results have to be proved by intra- and inter-laboratory check tests.

- *Cost/benefit evaluation* [7]: The manager has the task of analysing the cost–benefit of each analytical model of food analysis. After evaluation he has to find a compromise between the best analytical approach and economical feasibility.

## CONCLUSION

The introduction of quality assurance programmes in laboratories is a priority because:

- it improves the laboratory operations and the confidence in the results
- it gives the laboratory the possibility of receiving an accreditation with less difficulty.

The success of the application of a quality assurance programme in food analysis is principally dependent on good management, the motivation of experienced and well-trained analysts and the availability of good scientific documentation. In my opinion the question, 'Where is quality assurance in food analysis leading?' must necessarily be met with the answer, 'No laboratory can operate successfully without a quality assurance programme'.

## REFERENCES

[1] *Communication on the free movement of foodstuffs within the Community 89/ C271/03.* OJ No. C271/3, 10.24.1989.
[2] *Council Directive 83/189/EEC of 28 March 1983 laying down a procedure for the provision of information in the field of technical standards and regulations.* OJ No. L109, 04.16.1983, p. 8.
[3] *Koninklijk besluit betreffende de inrichting en de werking van de laboratoria voor ontleding van voedingswaren of -stoffen en andere produkten.* KB 23 October 1965.
[4] *Council Directive 88/320/EEC of 9 June 1988 and 90/18/EEC of 18 December 1989 concerning the inspection and verification of Good Laboratory Practices.* OJ No. L145/35, 06.11.1988 and OJ. No. L11/37, 01.13.1990.
[5] Garfield, F. M. (1984) *Quality Assurance Principles for Analytical Laboratories*, AOAC.
[6] Conacher, H. B. S. (1987) Importance of Quality Assurance in Canadian Pesticide Analysis, *J. Assoc. Off. Anal. Chem.* **70** (6) 941.
[7] Stephany, R. W. (1989) Quality Assurance and control in the analysis of foodstuffs, residue analysis in particular, *Belgian Journal of Food Chemistry and Biotechnology* **44** (4) 139.

# IV.3

## *Scire est decidere* (Knowing is deciding)

**Pieter L. Schuller**
Chief Inspectorate for Health Protection, PO Box 5406, 2280 HK Rijswijk,
The Netherlands
**Rainer W. Stephany**
National Institute of Public Health and Environmental Protection, PO Box 1,
3720 BA Bilthoven, The Netherlands

Although the origin of analytical chemistry can be traced back to ancient times, applied analytical chemistry was almost unknown at the beginning of the nineteenth century. By 1880, however, compositional analysis had developed to the point where it could provide a general knowledge of the proximate composition of most basic foods and had begun to develop useful and practical applications to everyday problems involving fraudulent substitution and adulteration.

For the most part, the public had faith in science and science was able to maintain that faith through developments which recognizably improved the lot of mankind. But now that faith has disappeared. How sure is science? Are we sure about our knowledge? That is the question behind the title of the paper, although the title in itself is given as a statement. The title is a play upon the words of the postulate *SCIRE EST MENSURARE* of Johannes Kepler, who lived from 1571 to 1630. *Scire est mensurare* is correctly translated as 'to know is to measure!' and not, as is often done nowadays, 'measuring is knowing'. Will merely measuring do us know? How do we know that what we are measuring is correct? Knowledge is subject to a constant process of enrichment and correction. By characterizing chemistry as an empirical science it becomes a base for judgement, for enrichment and for correction. Empiricism is represented by findings through observation. The ultimate question is then: how reliable are our *observations*? One aim of the physical sciences has been to give an exact picture of the material world. One achievement of physics in the twentieth century has been to prove that that aim is unattainable. What has

Note: Any opinions, findings, conclusions and recommendations expressed in this paper are those of the authors and do not necessarily reflect the views of the Chief Inspectorate for Health Protection or of the National Institute of Public Health and Environmental Protection.

been proved is that there is no absolute knowledge. All information is imperfect, one has to treat it with humility. That is the human condition and that is backed up by the theory of quantum physics. Looking for the truth, one needs to doubt each finding obtained through observation.

There is an analogy between what is called the march of physics and the progress of chemistry in so far as it concerns the gaining of detailed information. Since the seventeenth century, a substantial development has taken place in the ways we can make observations. Remember the construction of the microscope and the telescope. The development of new improved ways to observe has always in the past and even today induced enthusiasm; few expressed distrust about the position of mankind between two infinites, infinite space and ever-smaller-growing structures. Year by year, mankind developed and constructed more and more sophisticated instruments with which he could observe nature in finer detail. Yet when we look at our observations we see that they are still fuzzy and we feel that they are as uncertain as ever. Here we are face to face with the crucial paradox of knowledge. Irrespective of whether one looks at the position of a star as it was determined by classical astronomical instruments, or now by today's most sophisticated ones, or whether one compares the closeness of the results of fat determinations in food in 1890 or in 1990, one is astonished and chagrined to find that measurements suffer from discrepancies as much as ever. This is emphasized by Table 1, which summarizes part of the results of the Eurofood Inter-laboratory Trial 1985.

**Table 1** — Eurofood Inter-laboratory Trial 1985: Egg powder

|  | Dry weight | Protein | Fat | Total dietary fibre |
|---|---|---|---|---|
| Mean (%) | 95.26 | 52.983 | 37.779 | 0.361 |
| Range (%) | 94.6–96.5 | 49.7–56.9 | 29.4–44.2 | 0–0.8 |
| CV% | 0.54 | 3.1 | 8.9 | 117 |
| CV within | 0.18 | 1.4 | 2.0 | 22.6 |
| CV between | 0.50 | 2.8 | 8.7 | 115 |
| Confidence limit | 1.0 | 6.0 | 17.7 | 233 |

RIKILT Report 85/67.

The Eurofood subcommittee on laboratory analyses conducted, in 1985, a Eurofoods Inter-laboratory Trial with the aim of determining the influence of differences in analytical and other procedures between those laboratories that supply the data for nutrient values in food tables. One American and 19 European laboratories that regularly contribute nutrient values to nutrient databanks were invited to participate in the trial. Six foods were selected, among them egg powder.

The conclusion of the report was: 'Leading laboratories in different countries may produce different values for proximate constituents in common foods' [1].

Nowadays nanograms, picograms, femtograms or even attograms are determined. So, within 50 years, the limit of quantification has been lowered by a factor of approximately $10^{15}$. The distance between the Earth and the Moon has now been measured to within one nanometer! During the symposium 'The art of chemistry, where will it be forty years from now?' held at the 1989 Pittsburg Conference in Atlanta (USA) the expectation was expressed that the detection limits in forty years will approach brontograms. A brontogram equals $10^{-21}$ of a gram, so detection limits will approach Avogadro's number! The ability to detect and measure increasingly minute amounts of substances will act, in the case of environmental hygiene, including foodstuffs, to bias the perception of problems, by posing them in terms of highly refined measurement capabilities to the detriment of effective problem analysis. The pressures are exacerbated when the problems under consideration have substantial implications for public policy, particularly where political concerns are involved. One might say that the very success of analytical instrumentation has fostered new problems while leaving some old problems unsolved. This has caused T. Hirschfield to complain: 'Today's maximum analytical capacity is tomorrow's minimal legal requirement.'

One had hoped that human error would disappear and that one would have the Omnipotent's view. But it turns out that errors cannot be taken out of the observation, and that is true of all human activities. Bronowski's [2] conclusion was: 'Errors are inextricably bound up with the nature of human knowledge.'

Food legislation established, in general, two principal offences: the mixing of injurious ingredients; and selling to the prejudice of the purchaser, a food not of the nature, substance or quality demanded. The second of these requires evidence of composition. Therefore enforcement of food legislation is ultimately based on the ability of analysts to accurately identify and quantify (trade) levels and often trace levels of both inorganic and organic substances in various foods. Usually when one speaks of accuracy as such, one refers to a true value, in the sense that exactly that amount is present in a given matrix. In the light of the above, this is a tragic point of view. However, going back to the origin of the word 'accurate' from Latin 'accuratus', we find it means prepared with care, careful, exact in the sense of freedom from mistake or error. It would be a great task for all philosophical scientists to make it clear to judges and lawyers that the word 'accuracy' in the sense of 'prepared with care and free of mistakes' applies to the legal responsibilities of regulatory agencies involved with the detection, identification and measurement of particular substances in various matrices and not only to the analytical result involved.

It became clear to analytical chemists that, in general, the lower the content of the chemical one was looking for, the so-called analyte, the higher the unreliability of the analytical result. If one extends the measurable and detectable levels lower and lower, one meets a form of Heisenberg's principle: 'Measurements at the lower end of the scale are achievable only at the expense of reduced precision.'

From the results of a century of comparative quantitative inter-laboratory testing, conducted under the guidance of the Association of Official Analytical Chemists, Horwitz derived an empirical correlation between the content of the analyte and the coefficient of variation in the analytical result [3, 4]. He found an exponential increase of the coefficient of variation with decreasing content, typically ranging from 15 per cent at the part per million level to more than 40 per cent at the 1 part per billion level. This, in general, is what one has to expect; it is nothing to be ashamed of!

Notwithstanding this, all of society still expects from the chemist reliable analytical results upon which decisions have to be taken. This is only possible if the analytical chemist receives an analytical problem instead of only a sample. The analytical chemist himself is the only one who can deliver a complete answer (result, interpretation of the result, and conclusion therefrom).

The demand for reliable, sensitive, specific, fast, low-cost methods for residue analysis is rapidly growing, especially in the field of environmental protection, including food analysis [5]. Enforcement of food legislation methods should produce, in principle, no false negative results to protect the consumer and no false positive results to protect the producer. Such perfect methods do not exist, as pointed out above. So one has to accept false negative as well as false positive results, the question being only how many or, better, how few. That question cannot be answered by the analytical chemist but should be decided by policy makers in relation to the appropriate legislation and its impact on society. The analytical chemist needs guidance to tune his analytical strategy to the specific demands of his policy makers. It is possible to calculate the number of false positive results to be expected in a population of samples once policy makers have indicated an acceptable probability in percentage of false positive results and false negative results. The principal policy maker in his turn should realize that each strategic analytical model for residue control has its own price. A cost–benefit analysis has to be performed to compromise between (theoretical) best analytical approach and commercial feasibility. Table 2 summarizes some commercial prices for various analytical techniques.

**Table 2** — Estimate[a] in ECU for analysis of anabolic residues by commercial laboratories in The Netherlands

| Method | Number of analytes per sample | Cost | | Number of samples per run |
|---|---|---|---|---|
| | | per sample | per analyte | |
| RIA | 1 | 12–20 | 12–20 | 5–30 |
| TLC | 10 | 80–120 | 8–12 | 3–10 |
| GC–MS | 100 | 120–160 | 1.2–1.6 | 10 |

[a]Quantity-produced, status 1988.
RIA, radioimmunoassay; TLC, thin-layer chromatography; GC–MS, gas chromatography–mass spectroscopy.

Although the estimates are valid for the analysis of residues of anabolic agents in the meat of slaughter animals and doping agents in sports in continuous routine programmes, the prices given will not differ very much for other residue analyses [6].

If one estimates the cost of a single sample for an *ad hoc* identification of a single analyte an HPLC/RIA or Immunogram, it will come to around 120–200 ECU, and a HPLC/GC–MS run will cost about 360–440 ECU.

In order to achieve the certainty asked for, the usual practice up till now has been to describe an analytical method meticulously. A minute description of a method, however, is not enough. It is often forgotten that in order to obtain a good and acceptable analytical result one has to deal with the 'five-M-factors' concept. Men (technician and staff), Method (analytical procedure), Machine (technical equipment and reagents), Material (sample and standards) and Manipulation (data handling, interpretation): each of the factors contributes to the final unreliability of the analytical results, and all must have the attention they deserve! In the context of this paper, only the 'method' factor is emphasized.

Quantitative analysis is concerned with the methods of determining the relative proportion of constituents of a substance. The quality of quantitative methods is described by parameters like accuracy, precision, limit of detection, sensitivity. Only after a successful collaborative study or studies, is the method under consideration regarded nationally and internationally as validated. It has become clear to everybody working in the field that validation of methods for all possible situations along these lines is unrealistic because of the financial burden, the lack of analytical capacity and the impossibility to harmonize the use of highly sophisticated instruments and software between laboratories.

Qualitative analysis has for its objective the identification of a constituent of a substance. The quality of qualitative methods, for which the criterion of specificity is the most important one, cannot be evaluated in the terms mentioned. Yet a definite degree of certainty for qualitative methods is also needed, particularly for use as evidence in court. The first discussions on this item flared up in The Netherlands as a consequence of the fact that court cases on analytical control of veterinary drugs had to admit some false positive results. According to jurisprudential opinion, a probability of 95 per cent certainty is unacceptably low, but the demanding 100 per cent is judged unrealistic. Leaving the matter unresolved, an analytical strategy was developed in the case for control of diethylstilboestrol in bovine urine, consisting of a screening with a combination of column chromatography and off-line radioimmunoassay, confirmation by high-pressure liquid chromatography and off-line GC–high-resolution mass spectroscopy. This analytical strategy has proved to have an overall probability of a false positive result of less than 0.00001 per cent positive results (1 in $10^8$). When the tempest died down and common sense arose from the ruins, it was concluded that such an extreme analytical effort is far too expensive for common regulatory residue analysis. An error probability of 1:1000 or 1:10 000 seemed to be an acceptable and defendable compromise. The problem supplier and solver should agree *beforehand* about the various parameters related to residue analysis, e.g., required reliability, speed, costs, sample throughput, not forgetting the applicability of the results for the intended purpose [5].

To solve the problems connected with the validation of analytical methods, a group of international experts looked for another way out, and, as a result, an EEC Working Group came up with another approach for validation of analytical methods, *viz.* the concept of application of certain 'quality criteria'. As a result of a number of consultations during the period 1983–1989, the EEC published a decision of the Commission [7] concerning quality criteria for qualitative methods of analysis intended for use in chemical residue analysis. All criteria are focused to exclude false positive results [8, 9]. In the EEC document mentioned, quality requirements for identification of an analyte are presented for: gas chromatography with or without low-resolution or high-resolution mass spectroscopy, high-pressure liquid chromatography coupled with spectrometry, (high-performance) thin-layer chromatography with or without spectrometry.

Next the need was felt to rank the most commonly used methods according to a kind of degree of (un)certainty. An 'uncertainty' factor was introduced. The overall uncertainty factor of a method, $\Sigma_M$, can be regarded as the multiplication sum of the uncertainty factors of all parts of the method concerned:

$$\text{METHOD} = \text{ISOLATION} + \text{SEPARATION} + \text{DETECTION}$$
$$\Sigma_M \quad = \quad \Sigma_I \quad \cdot \quad \Sigma_S \quad \cdot \quad \Sigma_D$$

Firstly, attention was paid to the uncertainty of identification of a compound [10]. It was understood that the uncertainty factor has at least two components:

(a) the uncertainty about the position of the signal maximum, expressed in terms of retention time, $R_f$-value, wavelength and mass number and
(b) uncertainty concerning the signal width, which can be expressed as band width at half height and is related to the resolution.

For discussion, the uncertainty factors tabulated in Table 3 have been proposed [8].

**Table 3** — Uncertainty factors of method steps in identifying an analyte

| Step | $\Sigma$ |
| --- | --- |
| Immuno-assay | 1:2–5 |
| One-dimensional TLC | 1:20 |
| Two-dimensional TLC | 1:400 |
| HPLC | 1:8–50 |
| GC | 1:50–200 |
| UV–VIS (full spectrum) | 1:8–50 |
| Low-resolution MS (4 diagn.ions) | $:10^8$ |
| MS–MS | $\gg 1:10^8$ |
| Infrared | $1:10^{14}$ |

TLC, thin-layer chromatography; HPLC, high-pressure liquid chromatography; GC, gas chromatography; UV–VIS, ultraviolet–visible light; MS, mass spectroscopy.

Another Working Group called together by the EEC discussed the use of methods based upon molecular spectroscopy and concluded that for unambiguous identification of a substance, detailed information on the molecular structure of the analyte is essential [11]. The total information on the molecular structure of the analyte is the result of the sum of information derived from each of the various steps of a method of analysis. Analytical steps based upon molecular spectroscopy all provide *direct*, more-or-less detailed information on the structure of the analyte obtained by interaction of molecules and electromagnetic radiation. Incidentally, it should be realized that for the greater part of residue analysis in food control laboratories the selective analytical steps are based either on chromatographic techniques or progressively on more immuno affinity. These steps provide only more-or-less general indirect information obtained by interaction between various kinds of molecules. Table 4 lists the spectroscopic techniques used for identification purposes, in order of decreasing information content of the full spectrum. It is seen that the greatest amount of direct information is gained from infrared spectroscopy [12].

**Table 4** — Spectroscopic techniques listed
in descending order of direct structure
information obtained from full spectrum

| |
|---|
| Infrared |
| Nuclear magnetic resonance |
| Proton magnetic resonance |
| Mass spectroscopy |
| High-resolution mass spectroscopy |
| Mass-selective detection |
| Ulraviolet–visible light |

In order to extend the idea of the use of uncertainty factors, a group of experts from EC Member States was asked to give a 'desk-top' opinion on a set of twelve different commonly used combinations of qualitative selective analytical steps in residue analysis [13]. The experts were asked, on the basis of their long-established experience in analytical chemistry, to rank the methods looked at from low to high selectivity for the entire method. The results from the 25 experts asked were combined and then divided by the highest score to yield an *experimental* 'relative selectivity index'. From the *theoretical* point of view, each of the individual analytical steps used in the methods looked at were given a 'partial selectivity index' which was based upon, among other things, the uncertainty factors. As an example, the partial selectivity indices for the step of primary isolation from the sample material is given in the Table 5.

By summation of the partial selectivity indices the selectivity index of each of the methods looked at was calculated. Theoretical and experimental relative indices showed a very good correlation. The selectivity indices — which may be regarded as a combination of EC criteria and uncertainty factors, amplified and completed with

**Table 5** — Partial selectivity indices for primary isolation steps

| Analytical step | Index |
| --- | --- |
| Extraction (e.g. simple liquid–solid partition) | 0 |
| Specific extraction (e.g. ion-pair extraction) | 1 |
| Solid-phase extraction | 2 |
| Immuno-affinity chromatography | 3 |

the experience of a number of highly qualified laboratory experts — can be used, and are recommended to be used, to define the reliability of a method needed for a certain purpose. For different purposes, different demands are made of analytical strategies. Table 6 summarizes the guidelines for practical uses of the selectivity indices.

**Table 6** — Reliability of analytical methods classified according to selectivity index

| Purpose | Selectivity index of appropriate method |
| --- | --- |
| Within laboratory orientation | 2 |
| Screening | 3 |
| Surveillance | 3 |
| Between laboratory orientation | 4 |
| Confirmation | 7 |
| Forensic, national | 8 |
| Forensic, international | 10 |

**CONCLUSION**

In view of the lengthiness of the traditional procedure for arriving at fully harmonized methods to be used (inter)nationally in cases of enforcement or dispute, this paper draws attention to attempts made so far within the European Communities to develop a new yardstick to objectify comparison of methods. These attempts are based upon agreement concerning a concept of quality criteria and analytical reliability strategies supported by such criteria. The strength of this approach is that, among all involved official analytical chemists, consensus has been or should be achieved before any dispute or enforcement problems arise. The general idea behind the new concept is already widely accepted, or even mandatory, within the European Communities and is gaining more understanding within other international bodies, e.g. Codex Alimentarius, WHO and AOAC. However, many details and extensions still need further elaboration and discussion. All those interested or involved in regulatory official and/or forensic analytical chemistry, especially chemometrists, are invited to suggest improvements or to participate in open-minded discussions. The quality criteria approach, combined with an adequate quality assurance/quality control (QA/QC) strategy, will finally result in a flexible, feasible analytical strategy for reliable residue analysis.

## ACKNOWLEDGEMENT

The authors are indebted to Dr P. Brereton (MAFF, Norwich, UK) for the critical reading and correction of the manuscript.

## REFERENCES

[1] Hollman, P. C. H. & Katan, M. B. (1985) *Report 85/67*, State Institute for Quality Control of Agricultural Products, Wageningen, The Netherlands.

[2] Bronowski, J. (1973) *The Ascent of Man*, Brown and Company, Boston Toronto.

[3] Horwitz, W., Kamps, R. L., & Boyer, K. W. (1980) *J. Assoc. Off. Anal. Chem.* **63** 1344–1354.

[4] Boyer, K. W., Horwitz, W., & Albert, R. (1985) *Anal. Chem.* **57** 454–459.

[5] Schuller, P. L., Stephany, R. W., Egmond, H. P. van, & Vaessen, H. A. M. G. (1979) Applied Science Publishers, Barking UK, pp. 37–74.

[6] Stephany, R. W. (1989) *Belgian J. Fd. Chem. & Biotechn.* **44** 139–153.

[7] Anonymous (1987) *Off. J. Eur. Com.* L223:18–36 (doc 87/410 EEC).

[8] Ruig, W. G. de., Stephany, R. W., & Dijkstra, G. (1986) CEC doc VI/4705/86 Brussels.

[9] Ruig, W. G. de, Stephany, R. W., & Dijkstra, G. (1988) *J. Assoc. Off. Anal. Chem.* **72** 487–490.

[10] Ruig, W. G. de, Dijkstra, G., & Stephany, R. W. (1989) *Anal. Chim. Acta* **223** 277–282.

[11] Anonymous (1988) *Off. J. Eur. Com.* L351:39–50 (doc 89/610 EEC).

[12] Stephany, R. W. (1989) *J. Chromatog.* **489** 3–9.

[13] Ginkel, L. A. van, Stephany, R. W. (1990) *Berichten uit het RIVM 1989*, pp. 337–339.

# IV.4

# Summary report — Food control and analysis

**A. Ruiter**
Department of the Science of Food of Animal Origin, Faculty of Veterinary
Science, University of Utrecht, The Netherlands

For adequate food control it is necessary that suitable analytical methods are available and that we can guarantee the quality of these methods. Furthermore, we have to know which kind of information is provided by these methods and which decisions can be taken. All these aspects have been dealt with in this Part.

Professor Dr W. Baltes (Berlin) has presented an overview of fast analytical methods in food analysis. As food production should be monitored on a large scale, and because of the demands for an effective food control, these fast methods are increasingly important.

Fast methods can be defined in two ways: according to the equipment and according to the kind of method.

Equipment may be either low-cost (such as test kits) or high-cost (such as HPLC, NIR or NMR methods). High costs may be justified by the number and quality of data that can be produced in a short time.

A definition independent from the expenditure seems the most useful one. As for the kind of method, one can distinguish between:

  (i)  detection and screening methods;
 (ii)  test or methods of limited value;
(iii)  quantitative or semi-quantitative methods;
(iv)  methods for process control.

Fast methods can be introduced either by *substitution* or by *organization*. In the first case an existing method is simply replaced by a faster method. The organization-type introduction does not lead to a method which is essentially different from the original one but takes advantage of automation. One of the examples that can be given is continuous-flow analysis.

Apart from these, *immediate* methods in which physical properties are measured (e.g., NIR spectrometry) may give fast and valuable information. A very special development for fast food control is image analysis. This technique can be used for several purposes, e.g., to detect bones in fish fillets or to measure the fat/lean ratio in meat products. Electric and biochemical sensors were mentioned as well.

As for fast methods in the laboratory, illustrations were given with regard to the methodologic changes with respect to the analysis of fat, protein and solid matter.

Separation methods play a very important part in modern food analysis. Chromatographic techniques were reviewed and it was stressed that many of these can also be used as screening methods.

For polar compounds which cannot be easily determined by gas chromatography, capillary isotachophoresis has been proved very useful. It has the advantage that a time-consuming clean-up procedure is unnecessary. This technique is applied to inorganic anions, organic acids, vitamins, biogenic amines and artificial sweeteners.

Low-resolution NMR spectrometers have been on the market since about 1965. The principles of NMR were roughly explained, and the fact was stated that modern instruments can quantify fat and moisture contents within seconds. This technique is especially capable of measuring the portion of liquid matter in solids, e.g., the portion of liquid fat in solid fat.

Fast methods can also be based upon measurement by ion-selective electrodes. It was explained how these electrodes were derived from the glass electrode by which hydrogen ions are measured. Ion-selective electrodes are also suitable as detectors in flow injection analysis.

Finally it was pointed out that test kits may play an important role in modern food control. Test strips containing immobilized reagents are used for estimation of a number of compounds.

Both development and use of fast methods will greatly increase in the future; as for their accuracy, however, most of these methods will orientate themselves by classical standard methods.

Dr H. A. S. Beernaert (Brussels) has presented the second chapter, in which he discussed quality assurance in food analysis.

It is a matter of fact that, in order to guarantee the safety of foodstuffs, the quality of analysis must be assured. In the European Community two groups of standards have been prepared and adopted by the Council. The first one (EN 29000) specifies the criteria for the quality assurance (QA) of analytical services and systems. The criteria laid down in the other one (EN 45000) are those to which laboratories should conform and which should be used by accreditation bodies in accrediting laboratories, by public authorities and by any other organizations when designating laboratories.

Accreditation bodies are known in The Netherlands, the United Kingdom, France and Denmark. In Belgium, a National Council of Accreditation and certification has been installed. The Inspectorate is paying special attention to the integration of Good Laboratory Practice (GLP) in the quality system.

To assure the quality of food analysis in both official and private laboratories the staff has to develop the framework for planning the strategy, which is based on a 'five-forces' model, these being marketing, financial impact, investment, exploitation and results.

On the base of a graphical model, it was discussed by Beernaert how quality can be integrated in a laboratory. Essential elements of QA are: prevention (to ensure that analytical systems are working properly), assessment (which includes periodic checks) and correction (in order to determine quality defects and to restore proper functioning).

Critical control points are: sample collection (including preservation and shipment), sampling for analysis, methods of analysis, inadvertent errors, instrumentation, technicians, validation of results and cost/benefit analysis.

For a QA programme several types of methods can be distinguished, i.e., official methods, reference methods, screening methods, routine methods and automated methods. Their suitability has to be tested with the help of reference materials or inter-laboratory ring tests.

In conclusion, the priority of QA programmes in laboratories was stressed because of the improvement in laboratory operations, confidence in results and the possibility to receive accreditation with less difficulty. Success in applying QA is principally dependent on good management, the motivation of trained and experienced technicians, and good scientific documentation.

The final chapter, by P. L. Schuller and R. W. Stephany was, at least partially, a philosophic approach to analytical food chemistry.

By 1890, compositional analysis had developed to a point where it could provide a general knowledge of the proximate composition of most basic foods and where the foundation had been laid for methods to detect frauds and adulteration.

As the greater part of the public has faith in science, the results of inspection by means of scientific methods have been generally accepted. However, there is an ever-increasing number of individuals who raise the question whether we can always be sure about our knowledge. This, in fact, was the question behind the title of that chapter.

By characterizing chemistry as an empirical science and since empirism is represented by findings through observation, the ultimate question concerns the reliability of these findings. Looking for the truth, we need to doubt each observational finding, as everybody has his own experience of reality. Observational findings and experience of reality are largely dependent on the experimental technique. This also holds for analytical food chemistry, which was demonstrated in the chapter with an example regarding proximate constituents in common foods: in the report of a trial it was concluded that leading laboratories may produce different values for these compounds.

Another point of interest is that the lowering of detection limits, a trend still going on in analytical food chemistry, may lead to a gigantic hysteria about immeasurably small risks. The lower the limit of detection, the greater the gap between the result obtained and the knowledge of the consequences of these results. It should be realized that if we lower our detection levels we meet a form of Heisenberg's principle: 'Measurement at the lower end of the scale is achievable only at the expense of reduced precision.' A large problem is that today's maximum analytical capacity may be tomorrow's legal requirement.

It is a matter of fact that errors never can be excluded. However, we have the task of making it clear to judges and lawyers that *accuracy*, in the sense of 'prepared with care and free of mistakes', applies to the legal responsibility of the regulatory agencies involved in analysis. In this context, attention has to be paid to GLP as well as to QA.

New methods for detecting and quantitating adulteration would present a major advantage in public protection. Some governments, however, seem to adopt a policy of downgrading the economic fraud regulation in favour of regulatory functions focused upon health risks which, in many cases, are still to be proven.

For enforcement of food regulations, methods should produce no false negative results to protect the consumer and no false positive results to protect the producer. As such ideal methods do not exist, it was discussed how a workable compromise can be achieved. An error probability of 1:1000 to 1:10000 seems to be an acceptable compromise.

To obtain an acceptable result we have to deal with the so-called 'five-M concept', i.e., Men, Method, Machine, Material and Manipulation. All of these items must be given the attention they deserve.

Quality criteria were set up by an EC Working Group on Analytical Methods which were primarily focused on the exclusion of false positive results. At first, attention was paid to the uncertainty in identifying a compound. Another EC Working Group discussed the use of methods based upon molecular spectroscopy and concluded that detailed information about the molecular structure is essential. It should be realized that chromatographic and immunochemical methods provide only indirect information.

Finally, the selectivity of primary isolation steps was discussed.

The intention in Chapter IV.3 was to draw attention to attempts made so far within the EC to develop a new yardstick to objectify comparison of methods.

In discussion, the question arose as to who should establish the limits for unwanted compounds in food. It was stressed that the zero tolerance concept should be left aside, as it is impracticable and also untenable from a theoretical point of view.

It is the analytical chemist rather than the toxicologist who has to propose a level, whilst technological considerations also may contribute. Political discussions will depend on the wishes of the public, which makes the availability of good and objective information concerning possible risks all the more important. This, however, is not among the common tasks of the analytical chemist.

# Part V
## Safety assessment

# V.1

## Safety assessment: the process/procedure to be followed

**Carol J. Henry, Ph.D., D.A.B.T.**
ILSI Risk Science Institute, 1126 Sixteenth Street, NW, Washington, DC 20036, USA

### INTRODUCTION

The process of safety assessment as it is applied to food and food additives has changed dramatically during the last 40 years and is still in a dynamic phase. The origin of these changes can be traced in part to heightened awareness about the public's exposure to various agents suspected of causing serious disorders — cancer, birth defects, nervous disorders, other chronic diseases of unknown origin (Table 1). The real or suspected adverse effects from these agents have created health and safety concerns that did not exist to any degree before the 1950s (ILSI Risk Science Institute, 1987).

The political reaction to these trends has been the passing of food and drug, occupational safety and health, and environmental laws, which to varying degrees require regulation of these agents. Three United States agencies have authority for several different laws concerned with food: the Food and Drug Administration (FDA) has responsibility for food, drugs, cosmetics, food additives, colour additives, new drugs, animals and feed additives, and medical devices under the Food Drug and Cosmetic (FDC) Act, as amended; the Environmental Protection Agency (EPA) has responsibility for pesticides under the Federal Insecticide, Fungicide and Rodenticide (FIFR) Act of 1948, as amended, as well as for establishing tolerances for pesticide residues on raw commodities under Section 408 of the FDC Act (National Research Council, 1987); the United States Department of Agriculture (USDA) also has responsibility for food, feed, colour additives, and pesticide residues under the Poultry Products Inspection Act of 1968 (ENVIRON, 1988).

Typically, these laws have required that the principal federal regulatory agencies impose limits on the extent of human exposure to these agents. In deciding on the appropriate measures to restrict exposures, regulatory agencies typically assess the health risk that might result from particular use or exposure practices. *Risk*

**Table 1**— Some environmental agents that have caused public concern regarding real
or suspected adverse effects in the past 30 years

| Agent | Use or source | Concern |
|---|---|---|
| Acid rain | Burning fossil fuels | Environmental damage (trees, fish, etc.) |
| Aflatoxin | Mould toxin on peanuts and corn | Cancer |
| Agent Orange | Herbicide used in Vietnam | Cancer, birth defects, other health effects |
| Alcohol | Alcoholic drinks and medications | Birth defects, neurological impairments, cancer |
| Aldicarb | Pesticide | Toxicity from residues in watermelons |
| Aldrin, dieldrin | Insecticides | Cancer |
| Aluminium | Cooking pots | Alzheimer's disease |
| Amitrole | Herbicide used in cranberry fields | Cancer |
| Antibiotics in animal feed | Improving growth rate of food-producing animals | Increase in occurrence of drug-resistant pathogens |
| Asbestos | Fireproof insulation and ceiling tiles | Cancer |
| Aspirin | Pain killer, fever reduction | Reye's syndrome |
| Carbon black | Photocopy toner | Mutagenic contaminants |
| Carbon tetrachloride | Dry-cleaning fluid | Cancer, liver disease |
| Chlorine | Drinking water disinfection | Carcinogenic byproducts of chlorination |
| Chloroform | Toothpaste, mouthwash | Cancer |
| Cobalt | Foam stabilizer in beer | Cardiomyopathy |
| Cosmetic colours | Hair dyes, makeups, lipstick | Cancer |
| Cyclamate | Artificial sweetener | Cancer |
| DDT | Insecticide | Cancer, environmental damage (eggshell thinning) |
| DES | Drug to prevent miscarriage | Birth defects, cancer |
| Dioxin | Contaminant in herbicides, combustion products | Cancer, birth defects, other health effects |
| EDB | Fumigant for grain | Cancer |
| Fluorocarbon propellants | Aerosol spray cans | Depletion of stratospheric ozone layer |
| Formaldehyde | Urea-formaldehyde foam insulation | Cancer, irritation of eyes and respiratory system |
| Heptachlor, chlordane | Insecticides | Cancer |
| Hexachlorophene | Disinfectant | Neurological effects in infants |
| Kepone | Insecticide | Sterility, environmental effects |
| Lead | Old paint, automobile exhaust | Neurologic and behaviour disorders in children, heart disease in adult males, reproductive effects |
| Mercury (organic) | Fish and shelfish | Neurological damage |
| Methylene chloride | Decaffeinated coffee, paint stripper | Cancer |
| Polybrominated biphenyls | Fire retardant mislabelled | Death of cattle, illness of humans |
| Polychlorinated biphenyls | Cooling fluid in transformers and capacitors | Cancer, other health effects, environmental effects |
| Perchloroethylene | Dry-cleaning fluid | Cancer, liver damage |
| Phenacetin | Pain reliever | Kidney damage |
| Phosphate | Phosphate-containing detergents | Environmental effects |
| Radiation/radon | Bricks containing radioactive soil, etc. | Cancer |
| Saccharin | Artificial sweetener | Cancer |
| 2,4-D | Herbicide | Cancer |
| 2,4,5-T | Herbicide | Birth defects, cancer |
| Trichloroethylene | Degreasing solvent | Cancer |
| Thalidomide | Sedative | Birth defects |
| Tobacco smoke | Cigarettes, cigars | Cancer, heart disease, lung disease, etc. |
| Tris-BP | Flame retardant in children's sleepwear | Cancer |

Adapted from *Review of Research Activities to Improve Risk Assessment for Carcinogens*, ILSI Risk Science Institute (1987).

*assessment* is the term given to the process of assessing the health risk. For food and food additives, it is the tool used to evaluate the safety of such substances (Rodricks and Taylor, 1983).

Under Section 409 of the FDC Act, pesticide residues must be proven 'safe', which is defined as a 'reasonable certainty' that no harm to consumers will result when the additive is put to its intended use (National Research Council, 1987). In its common usage, 'safe' means 'without risk'. In technical terms, however, this common usage is misleading because science cannot ascertain the conditions under which a given chemical exposure is likely to be absolutely without a risk of any type. Such a condition — zero risk — is simply immeasurable. Science can, however, describe the conditions under which risks are so low that they would generally be considered to be of no practical consequence to persons in a population. As a technical matter, the safety of chemical substances — whether in food, drinking water, air, or the workplace — has always been defined as a condition of exposure under which there is a 'practical certainty' that no harm will result in exposed individuals. It should be noted that most 'safe' exposure levels established in this way are probably risk-free, but science simply has no tools to prove the existence of what is essentially a negative condition (ENVIRON, 1988).

One of the first principles in evaluating the 'safety' of substances is that substances cannot be classified simply as 'safe' or 'unsafe' (or as 'toxic' and 'non-toxic'). This type of classification, while common, is highly problematic and mislead-ing. All substances, even those which we consume in high amounts everyday, can be made to produce a toxic response under some conditions of exposure. As Paracelsus observed over 400 years ago, 'All substances are poisons; there is none which is not a poison. The right dose differentiates a poison and a remedy' (Klaassen *et al.*, 1986). The important question is not simply that of toxicity, but rather that of risk — i.e., what is the probability that the toxic properties of a chemical will be realized under actual or anticipated conditions of human exposure? To answer this question requires far more extensive data and evaluation than the characterization of toxicity. Risk assessment is the tool used to provide the answer to the question.

## THE ELEMENTS OF RISK ASSESSMENT

Risk is the probability of injury, disease, or death under specific circumstances. It may be expressed in quantitative terms, taking values from zero (certainty that harm will not occur) to one (certainty that it will). In many cases risk can only be described qualitatively, as 'high', 'low', or 'trivial'. A discussion of concepts and applications concerning risk was recently published and provides a good introduction for those new to the field (Glickman & Gough, 1990).

As shown in Table 2, all human activities carry some degree of risk. Many risks are known with a relatively high degree of accuracy, because we have collected data on their historical occurrence. For example, the number of deaths in the United States caused by motor vehicle accidents in one year is divided by the total number of people at risk (considered to be the entire US population) to give an individual risk for death in a given year of 1/4500. Lifetime risk assumes a 70-year lifetime to give a 1/65 probability of dying in a car accident over an entire lifetime.

**Table 2**— Annual risk of death from selected common human activities[a]

|  | Number of deaths[b] | Individual risk per year | Lifetime risk[c] |
|---|---|---|---|
| Coal mining |  |  |  |
|    Accident | 180 | $1.3 \times 10^{-3}$ or 1/770 | 1/17 |
|    Black lung disease | 1 135 | $8.0 \times 10^{-3}$ or 1/135 | 1/3 |
| Motor vehicle | 46 000 | $2.2 \times 10^{-4}$ or 1/4 500 | 1/65 |
| Truck driving | 400 | $10^{-4}$ or 1/10 000 | 1/222 |
| Falls | 16 339 | $7.7 \times 10^{-5}$ or 1/13 000 | 1/186 |
| Home accidents | 25 000 | $1.2 \times 10^{-5}$ or 1/83 000 | 1/1190 |

[a]Selected from Hutt (1978).
[b]Per representative year.
[c]Estimated upon a 70-year lifetime or 45-year work exposure.

The risks associated with many other activities, including the exposure to various chemical substances, cannot be readily assessed or quantified. Although there are considerable historical data on the risks of some types of chemical exposures (e.g., the annual risks of death from intentional overdoses or accidental exposures to drugs, pesticides, and industrial chemicals), such data are generally restricted to those situations in which a single, very high exposure resulted in an immediately observable form of injury, thus leaving little doubt about causation. Assessment of the risks of levels of chemical exposure that do not cause immediately observable forms of injury or disease (or only minor forms such as transient eye or skin irritation) is far more complex.

It is this type of risk assessment that is of great concern to the scientific and regulatory community and, just as importantly, to the general public. For industries associated with any type of chemical, food additive, or drug, the health risk assessment is the most critical activity associated with the substance.

The elements of risk assessment were described in the US in a report published by the National Academy of Science (NAS) in 1983, entitled *Risk Assessment in the Federal Government: Managing the Process*. In this report, research, risk assessment, and risk management were shown to interact and be related, but the report emphasized the separation between the scientific exercise of risk assessment and the policy exercise of risk management. From the NAS report, four elements were described in the risk assessment process: hazard identification, dose–response assessment, exposure assessment, and risk characterization, with recommendations and examples given for the type of scientific information needed for each element.

Risk assessment was defined as the process of assessing the possible adverse health effects in humans resulting from exposure to chemicals or other potential hazards (National Research Council, 1983). A brief review of the four steps follows.

**Hazard identification**
The first step of hazard identification involves gathering and evaluating data on the types of health injury or disease that may be produced by a chemical and on the

condition of exposure under which injury or disease is produced. It addresses the question, 'Does the agent cause the adverse effect?' The information concerning adverse effects may be found in a variety of studies, including epidemiology, animal toxicology, bioassay studies, human clinical studies, or *in vitro* studies.

Hazard identification may also involve characterization of the behaviour of a chemical within the body and its interactions with the body and with organs, cells, or even parts of cells. Such data may be of value in answering the ultimate question of whether the forms of toxicity known to be produced by a substance in one population group or an experimental condition are also likely to be produced in humans. Hazard identification can be considered as a qualitative risk assessment, simply determining whether and to what degree it is scientifically correct to infer that toxic effects observed in one setting will occur in other settings — e.g., whether substances found to be carcinogenic or teratogenic in experimental animals are likely to have the same result in humans who have been adequately exposed (ENVIRON, 1988).

### Dose–response assessment

The dose–response assessment addresses the question, 'What is the relationship between dose and incidence in humans?' It involves describing the quantitative relationship between the amount of exposure to a substance and the extent of toxic injury or disease. Data are derived from animal studies, or less often, from studies in exposed human populations. There may be many different dose–response relationships for a substance if it produces different toxic effects (e.g., cancer, birth defects) under different conditions of exposure (e.g., single compared to repeated exposures).

Generally, dose–response evaluation involves two extrapolations: one for interspecies adjustments for differences in size, lifespan, and basal metabolic rate; and one from high doses administered to animals to lower doses to which humans are likely to be exposed. The dose–response assessment should describe and justify the methods of extrapolations used to predict incidence and should characterize the statistical and biological uncertainties in these methods (National Research Council, 1983).

### Human exposure assessment

The exposure assessment addresses the question 'What exposures are currently experienced or anticipated under different conditions?' It involves determining the concentration of the chemical agent to which humans are exposed, the nature and size of the population exposed to a substance, and the magnitude and duration of exposure. Exposure refers to contact between the animal or human and the environmental media (air, water, diet, and soil) containing the substance. The evaluation could concern past or current exposures, or exposures anticipated in the future. The exposure assessment should describe and justify the methods of measurement, as well as characterizing the assumptions and uncertainties associated with the exposed population.

**Risk characterization**

Risk characterization generally involves the integration of the data and analysis of the first three components of risk assessment to determine the likelihood that humans will experience any of the various forms of toxicity associated with exposure to a substance. It addresses the question, 'What is the estimated incidence and severity of the adverse effect?' It is in this step of characterizing the risks, that the major assumptions, scientific judgement and uncertainties should be identified so that the risk estimate can be better understood. How the risk is characterized can be a critical link between the scientific risk assessment and the decision-making and communication issues that must be addressed in the risk management process. How these issues can be presented has been discussed in a recent publication developed by a committee of government and private sector scientists (American Industrial Health Council, 1989).

## LIMITATIONS AND ASSUMPTIONS

As has been discussed, risk assessment is a process. It provides a framework for evaluating data and presenting it to decision makers. However, risk assessment is limited by:

- lack of data about chemicals and adverse health effects;
- uncertainty in disease causation;
- uncertainty in extrapolating human risk from animal data.

These are significant uncertainties and limitations, which have resulted in the use of a set of assumptions or default positions. The assumptions or the uncertainties which abound in the risk assessment process have generated much controversy. However, in the absence of data or in the presence of uncertainty, public health officials tend to use assumptions that will not underestimate risk (Flamm, 1989; Park, 1989). The nine most generally agreed upon common assumptions in risk assessment are (US Office of Science and Technology Policy, 1985):

- In the absence of adequate human data, adverse effects in experimental animals are regarded as indicative of adverse effects in humans.

- Dose–response models can be extrapolated outside the range of experimental observations to yield estimates or estimated upper bounds on 'low-dose' risk.

- Observed experimental results can be extrapolated across species by an appropriately chosen standardized dosage scale.

- No threshold dose for carcinogenesis exists, although threshold levels may apply for other toxicologic outcomes.

- Average doses give a reasonable measure of exposure when dose rates are not constant in time.

- In the absence of pharmacokinetic data, the effective or target dose is assumed to be proportional to the administered dose.

- The risks from multiple exposure and multiple sources of exposure to the same chemical are usually assumed to be additive.

- Regardless of the route of exposure, 100 per cent absorption across species is assumed in the absence of specific evidence to the contrary.
- Results associated with a specific route of exposure are potentially relevant for other routes of exposure.

## CANCER VERSUS NON-CANCER RISKS

The risks of a substance cannot be ascertained with any degreee of confidence unless dose–response relationships are quantified. In the United States, a distinction between substances that cause cancer and those that do not has a major impact on the methods used to extrapolate human risks from animal data. Carcinogens are considered by regulatory agencies to pose a finite risk at all doses, while for non-carcinogens it is assumed there is a threshold dose below which no adverse effect is observed. This distinction results in characterizing the risk differently for these two classes of substances.

For non-carcinogens, a threshold dose or level of exposure is assumed below which no effect is observed. The dose–response evaluation involves estimation of the threshold dose and the determination of the no-observed-effect level (NOEL) from observations in exposed people or experimental animals. The acceptable daily intake (ADI) is estimated by dividing the NOEL by a safety or uncertainty factor. The ADI is compared to a hypothetical human maximum daily intake and if the maximum daily intake is smaller than the ADI, then no risk is assumed for the population. A critical area in this estimate is the magnitude of the safety or uncertainty factor, which can range from 10 to 10 000, based on the quality and quantity of data and on the policy of different regulatory organizations. For example, the Center for Food Safety and Applied Nutrition at the FDA uses safety factors between 100 and 2000, depending on the availability and type of data for analysis. The EPA uses the term 'reference dose' (RfD) in place of the term ADI to avoid the implied value judgement in the latter term that the dose calculated in this way is 'acceptable' (ENVIRON, 1988). The safety factors EPA uses are between 10 and 10 000 (Dourson & Stara, 1983). It has recently been suggested that the terminology associated with 'safety factors' or 'margin of safety' is inappropriate, and that alternative terms such as 'margin of protection' or 'protection factor' more accurately convey the aim and outcome of the process (Goldstein, 1990).

Under Section 409 of the FDC Act, the Delaney Clause prohibits the use of food additives found to induce cancer in man or animals. In contrast to the general safety standard for non-carcinogens, which recognizes the impossibility of assessing the complete absence of risk, the Delaney Clause has been interpreted as taking a 'zero risk' approach to substances implicated as carcinogens (Rodricks & Taylor, 1983). It should be stressed that this clause was enacted during a period when relatively few carcinogens had been identified and even fewer were believed to be present or associated with food.

The result in the United States of the interpretation of the Delaney Clause was an assumption that there was no threshold dose for carcinogens and that oncogenic risk was assumed to be a function of cumulative lifetime dose. Since most of the information about whether a substance is found to induce cancer is obtained from animal studies at high doses, scientists developed mathematical models to extrapolate the risk from the high dose levels used in the animal studies to determine the risk from the low dose levels to which humans would be potentially exposed. This modelling and extrapolation process has been termed 'quantitative risk assessment for chemical carcinogens'.

### Mathematical modelling

The modelling and extrapolation process is considered by many to be the single most important source of uncertainty in estimating risks from carcinogens (Samuels & Adamson, 1985; Flamm, 1989). The quantitative estimate of the risk at a particular low dose level is highly dependent on the mathematical form of the presumed dose–response relationship. Differences among models of at least 3 to 5 orders of magnitude are not uncommon (Samuels & Adamson, 1985; Whittemore *et al.*, 1986).

An example of the differences in the approach used by different regulatoryagencies for assessing risk can be demonstrated by review of the acceptable exposure levels set for 2,3,7,8-tetrachlorodibenzodioxin or TCDD (Table 3). While TCDD

**Table 3**— Acceptable Daily Intakes for TCDD proposed or adopted by various regulatory agencies[a]

| Agency | Dose–response extrapolation | Allowable intake (fg/kg day) |
|---|---|---|
| USEPA[b] | Linearized multistage | 6.4 |
| CDC[c] | Linearized multistage | 28–1428 |
| OME[d] | Safety factor (100) | 10 000 |
| SINH[e] | Safety factor (250) | 4000 |
| FEA[f] | Safety factor (100–1000) | 1000–10 000 |
| FDA[g] | Safety factor (77) | 13 000 |
| NYSDH[h] | Safety factor (500) | 2000 |

[a]From Paustenbach (1989).
[b]US Environmental Protection Agency (1985).
[c]Centers for Disease Control (Kimbrough *et al.*, 1984).
[d]Ontario Ministry of Environment (1985).
[e]State Institute of National Health, Netherlands (1982).
[f]Federal Environmental Agency, West Germany (1984).
[g]US Food and Drug Administration (USFDA, 1983; USEPA, 1984).
[h]New York State Department of Health (1983).

has been shown to be extremely toxic to some rodents, its carcinogenic potential has been the subject of considerable scientific controversy (Paustenbach, 1989). Whether it is considered to be a carcinogen determines if an acceptable exposure level is established. Generally an ADI is not derived from a NOEL for substances found to be carcinogenic. However, several agencies have found it appropriate to use a NOEL to calculate an acceptable exposure level for TCDD (Table 3). Using a range of safety factors as well as a cancer dose-modelling approach yields a 2000-fold range of acceptable exposure levels for TCDD (Paustenbach, 1989). This issue of how to interpret the biologic activity fo TCDD in rodent studies for human risk has captured much attention in the US and has been discussed in detail in the Regulatory Program of the United States Government (US Office of Management and Budget, 1990).

## Mechanistic information

From a scientific standpoint, substantial progress has been and is being made in understanding the mechanisms of toxicity and carcinogenesis, and in establishing causal relationships upon which safety assessments are made. Increasingly and particularly in Europe and Canada, it is recognized that distinctions can be made among carcinogens, based upon the differing mechanisms by which they act. Some substances directly initiate cancer and others are only secondarily involved in the process of carcinogenesis. Thus for some carcinogens, as for non-carcinogens, there are levels of exposure for which the possibility of harm to humans can be ruled out with reasonable certainty and for which an ADI-type or RfD approach to safety evaluation might be appropriate. Cohen & Ellwein (1990) recently presented evidence for the importance of mechanistic information in determining the existence of a threshold for the proliferative (and carcinogenic) response of non-genotoxic chemicals and the estimation of risk for human exposure. This area is under intense scrutiny and review, and will hopefully be clarified in the near future, particularly in the United States (US Environmental Protection Agency, 1988).

## Characterization of risk

The importance of the risk characterization step can be seen in interpretation of the concept 'one cancer in a million risk'. This is often the basis of regulatory decisions in the US. Substances with cancer risk estimates greater than one in one million ($10^{-6}$) have not generally been approved at the federal level (Rodricks et al., 1987).

The concept was explained in 1987 by the Commissioner of FDA, when he discussed the cancer risk from residues of methylene chloride, which is a solvent used to decaffeinate coffee:

'The risk of one in a million is often misunderstood by the public and the media. It is not an actual risk, i.e. we do not expect one out of every million people to get cancer if they drink decaffeinated coffee. Rather, it is a mathematical risk, based on scientific assumptions used in risk assessment. When FDA uses the risk level of one in one million, it is confident that the risk to humans is virtually nonexistent.' (Young, 1987)

Thus it can be seen that how the risk is characterized can make a significant impact on how the risk assessment is interpreted and received. These aspects of the perception and communication about risk have been addressed in a report entitled *Improving Risk Communication* (National Research Council, 1989). This report recommends that more efforts be expended in the interpretation of risk assessments and for communicating the technical points in ways that decision makers can use and the public can understand.

**Table 4**— Issues in health risk analysis

| Risk assessment issues | |
| --- | --- |
| Hazard identification | Use of animal data |
| | Negative epidemiological studies |
| Dose–response evaluation | Extrapolating from high dose to low dose |
| | Extrapolating from animals to humans |
| Humans exposure evaluation | Modelling vs. ambient monitoring vs. biological monitoring |
| Risk characterization | Qualitative or quantitative |

| Risk management issues | |
| --- | --- |
| Risk characterization | Quantitative vs. qualitative |
| Statutory and legal factors | Differing decision criteria |
| | Uncertainty in intent and interpretation |
| Economic and social factors | Quality and sources of information to be considered |
| Public concern | Fear of harm |
| | Technical understanding |
| | Involvement in decisions |

**CONCLUSION**

In summary, critical issues associated with risk assessment have been presented (Table 4). Extrapolation methods and the uncertainties associated with such methods play a major role in estimating risk, as does the qualitative or quantitative characterization of the risk. These risk assessment issues are considered in the risk management issues (Table 4), with characterization of risk being of concern to both. Throughout the risk management process, regardless of the agency or organization, decisions affecting risk and safety are made under varying degrees of uncertainty. The risk assessment/risk management paradigm is a critical and useful tool, providing a framework in which to harmonize and attempt to achieve consistency for complex and diverse scientific regulatory issues.

# REFERENCES

American Industrial Health Council (AIHC) (1989) *An Ad Hoc Study Group Presentation of Risk Assessments of Carcinogens.* Washington, DC.

Cohen, M. & Ellwein, L. B. (1990) Cell Proliferation in Carcinogenesis. *Science* **249** 1007–1011.

Dourson, M. L., & Stara, J. F. (1983) Regulatory History and Experimental Support of Uncertainty (Safety) Factors. *Regul. Toxicol. Pharmacol.* **3** 224–238.

ENVIRON Corporation (1988) *Elements of Toxicology and Chemical Risk Assessment.* Washington, DC.

Flamm, G. A. (1989) Critical Assessment of Carcinogenic Risk Policy. *Regul. Toxicol. Pharmacol.* **9** 216–224.

Glickman, T. S. & Gough, M. (1990) *Readings in Risk.* Resources for the Future, Washington, DC.

Goldstein, B. D. (1990) The Problem with the Margin of Safety: Toward the Concept of Protection. *Risk Analysis* **10** 7–10.

Hutt, P. B. (1978) *Food Drug Cosmetic Law J.* **33** 558–589.

ILSI Risk Science Institute (ILSI-RSI) (1987) *Review of Research Activities to Improve Risk Assessment for Carcinogens.* Washington, DC.

Klaassen, C. D., Amdur, M. O. & Doull, J. D. (1986). *Casarett and Doull's Toxicology: The Basic Science of Poisons.* Macmillan Publishing Company, New York, NY.

National Research Council (NRC) (1983) *Risk Assessment in the Federal Government: Managing the Process.* National Academy Press, Washington, DC.

National Research Council (NRC) (1987) *Regulating Pesticides in Food.* National Academy Press, Washington, DC.

National Research Council (NRC) (1989) *Improving Risk Communication.* National Academy Press, Washington, DC.

Park, C. W. (1989) Mathematical Models in Quantitative Assessment of Carcinogenic Risk. *Regul. Toxicol. Pharmacol.* **9** 236–243.

Paustenbach, D. C. (1989) *The Risk Assessment of Environmental and Human Health Hazards: A Textbook of Case Studies.* John Wiley & Sons, New York.

Rodricks, J. & Taylor, M. R. (1983) Application of Risk Assessment to Food Safety Decision Making. *Regul. Toxicol. Pharmacol.* **3** 275–307.

Rodricks, J. V., Brett, S. M., & Wrenn, G. L. (1987) Significant Risk Decisions in Federal Regulatory Agencies. *Regul. Toxicol. Pharmacol.* **7** 307–320.

Samuels, S. W. & Adamson, R. H. (1985) Quantitative Risk Assessment Report of the Subcommittee on Environmental Carcinogenesis, National Cancer Advisory Board. *J. Natl. Cancer Institute* **74** 945–951.

US Environmental Protection Agency (1988) Intent to Review Guidelines for Carcinogen Risk Assessment. *Federal Register* **53** 32656.

US Office of Management and Budget (1990) *The Regulatory Program of the United States Government April 1, 1990–March 31, 1991.* Office of Information and Regulatory Affairs, Washington, DC.

US Office of Science and Technology Policy (OSTP) (1985) Chemical Carcinogens: Review of the Science and its Associated Principles. *Federal Register* **50** 10372–10442.

Whittemore, A. S., Groffer, S. C., & Silvers, A. (1986) Pharmacokinetics in Low-dose Extrapolation Using Animal Cancer Data. *Fund. Appl. Toxicol.* **7** 183–190.

Young, F. (1987). Risk Assessment: The Convergence of Science and the Law. *Regul. Toxicol. Pharmacol.* **7** 179–184.

# V.2

## Safety assessment versus risk perception

**U. Versteegen** and **A. Warne**
CIBA-GEIGY Ltd, Basel, Switzerland

### RISK AND SAFETY REVISITED

The need to re-evaluate our thinking in the domain of safety assessment is rooted in the fact that, in the industrialized world, society no longer has to struggle for survival. Science has, in the main, tamed nature. We have found ways of ensuring an adequate supply of food, and ways of protecting ourselves against disease or neutralizing it. Our expectations of safety have increased, and our tolerance for risk has decreased. Finally we have the leisure, from our position of relative comfort, to take a look at the world that supplies us with all our raw materials.

It is through this process of observation that we have come to realize that the world's resources are finite, and should therefore be used carefully. It has also become clear that the innovations, previously welcomed for their benefits, are not without risks. The latter, however, only become apparent when the innovation is accepted into daily use. While science may have taken the sting out of nature, it appears that man has replaced it with risks of his own making.

Among the factors [1] influencing the change in society's attitude to risk appear the following three: an increase in innovation and the information that accompanies it; an increasing awareness of risk and the varying perceptions of it; and the democratization of the decision-making process with regard to the acceptability of risk. The rapid expansion of technological innovation has led to a worldwide information explosion. Approximately every five years, the sum total of knowledge is doubled, half of which is out of date only three or four years later. A new medical insight is made every five minutes, a new link in physics is discovered every three, and a new chemical formula every 60 seconds [2].

Suddenly more information than ever before is available to more people, regardless of their ability to absorb it. Geographical location, scientific expertise, and competence in decision making are no longer limiting factors of the accessibility of information. This overload of information has in turn led to anxiety, as the individual is unable to put this information in order of priority, to evaluate its importance, nor to interpret its personal meaning. People find they are not in a position to process information systematically, and are further hampered by contradictory opinions supplied by experts. Observation reveals that, ultimately, the public make decisions on the basis of perceived danger, and not on the basis of risk seen in terms of probabilistic data [3].

A second force of change is apparent in the heightened sense of risk awareness that emerges as evidence of risk and damage mount up. The effects of acid rain and CFCs on the environment, the fears raised by the accident at Chernobyl, the discovery of possibly life-threatening or crippling side-effects of pharmaceutical drugs, and the dangers of unintended additives in food have all shaken the public's faith in authority.

Traditionally uninvolved sectors of society are beginning to question the authority of scientific experts concerning the absence of psychological and socio-cultural aspects in decision-making. The classical triad of decision makers, comprising industry, regulators and scientists, is opening up to allow new groups a voice in the discussion of questions that encompass the implicit risk/benefit trade-offs that have to be made in matters relating to their well-being. The field is broadening to accept the views of experts representing the public and expressing their concerns. Pressure groups and consumer organizations are forming to deal with specific problems, and they have access to information on a worldwide basis from which to develop their arguments. Politicians are having to acknowledge the needs of the consumer, for whom the quality of life factor plays a significant role.

The main organ of communication and feedback is the mass media. With the advent of new technology, journalists now have access to more information more quickly and from all over the world. They are able to report events almost as they are happening. As a result, there is a growing awareness of differing viewpoints and priorities when talking about safety, and a growing realization that its definition depends on how a given situation is perceived. We are beginning to see that the same situation in which previously only benefits were perceived, must now be looked at in a different light. In this context, science has lost the privileged place it previously enjoyed in society. It has had to abandon its monolithic approach, in favour of one which undergoes a constant re-examination of its validity [4]. It must be accepted that no innovation is entirely safe and that unwanted secondary effects are inevitable. But if total safety is unrealistic, we can at least work towards broadening the margins of protection [5] and thus minimizing risk. The prerequisites necessary to achieve this are the understanding that risk and benefit cannot be considered either in absolute terms or in isolation, and that safety is a function of the balance between risk and benefit.

## DIFFERING APPROACHES TO RISK PERCEPTION AND SAFETY ASSESSMENT: A HIGHLY COMPLEX INTERACTION

The path to a higher level of safety, therefore lies in improving our understanding of the relationship betwen risk and benefit, and how they are perceived and assessed by the various constituencies. The way assessment and perception interrelate underpins the way our beliefs are formed. 'Assessment' is often thought of as an objective analytical process employed by experts, while 'perception' is frequently seen as an emotional approach that is subjective and largely uninformed.

This view, however, is misleading. Assessment and perception are complementary activities, one constantly fuelling the other. Each group involved in risk management — regulators, manufacturers, consumers, media and scientists — have a different perception of risk.

Not only does perception vary from group to group, but correspondingly the techniques used in assessment, too. Depending on the group, assessment may be biased towards the technical, political, socio-economic, psychological, social or cultural. The perception of risk is determined by the needs of each group. The commercial needs of a manufacturer differ substantially from the personal needs of a consumer. How groups perceive risk, influences how they assess safety, and this in turn lays the foundations for future perception and assessment, ultimately shaping and forming beliefs. Definitions of risk are similarly variable. From a technical standpoint, the definition of risk is the likelihood of an event combined with the severity of its outcome. But risk can also be seen as the estimate of damage and the cost expected from protective investments. It could equally well be defined as the relationship between the extent of possible control over an event and familiarity with that event. Comparative studies investigating definitions of risk have found that more than 30 exist [6]. The existence of this variety of ways of perceiving risk indicates that risk, as an absolute, does not exist — it is a social construct.

The matrix of social risk assessment (Fig. 1) combines perceptions with differing approaches to assessment and shows in broad terms how each group positions itself and carries out assessment according to the factors personally deemed important.

**Perception**

| Assessment | Manufacturers | Regulators | Scientists | Media | Consumers |
|---|---|---|---|---|---|
| Technical | ● | ● | ● | | |
| Socio-economic | ○ | ○ | ● | | |
| Cultural/ psychological/ social | | | ● | ● | ● |

Fig. 1 — Risk: the individual's or group's perception of what is important, and how this data is assessed. ● = traditional assessment approaches. ○ = evolving assessment approaches.

The only group to look at assessment in all of the three areas specified are scientists, while the positions of the media and consumers are diametrically opposed to those of manufacturers and regulators. The latter two groups are tentatively opening up their assessment procedure to encompass socio-economic factors in a systematic way; but psychological, societal and cultural aspects of safety assessment have not yet been considered.

The matrix shows that there is a large variance in standpoints regarding risk evaluation, and the positions of the constituencies are scattered. As each of the groups involved in risk management does not exist in isolation, it is logical to assume that their position relative to each other on issues of risk management will vary depending on each other's views and assessments. Little is known about the way in which different groups affect each other in defining their respective positions, and the consequent effect this has on their views on risk. As there is no universal procedure of risk perception and safety assessment, the matrix also shows that the subjective/objective approach to risk management is inapplicable.

### Risk perception by the public
Each group's views are shaped not only by the issue at stake but by its needs and environment. Where the public are concerned, there are four factors that strongly influence how risk is perceived and evaluated.

The *psychological* factor refers to the practice of overestimating the risks of dramatic but rare causes of death, such as airplane crashes, and underestimating those of familiar and frequent causes of death, such as car accidents. The more the risk is unknown, exotic, diffuse, less visible, involuntary and controlled by others, the greater the mistrust, criticism and rejection of it [7a,b].

In order to make a valid safety assessment, the public has to understand and interpret probabilistic information. However, most people have difficulty in doing so, particularly when the probabilities are small and the risks unfamiliar — the *communication* factor refers to the degree to which probabilistic information is understandable, thus enabling informed choice and rational utilization of innovation.

The *normative* factor refers to the values and norms brought into play when assessing risk information. These vary according to the cultural/social context of the decision-making groups within which the peer group influences the attitude of the individual towards risk assessment. Lastly, people interpret risk within the context of their personal experience and belief systems. An individual's experience of a car crash might have an effect on his or her family and friends regarding their perception of the relative safety of car travel. This is termed the *situational* factor.

All the above factors influence how risk is perceived and whether its significance will be attenuated or amplified. If an individual overestimates the risk involved in dealing with a problem, this will increase the real problem to the same extent as when he or she underestimates the seriousness of the risk. For example, if a patient decided that the risk of suffering side-effects when taking specific medication is too great, and consequently stops taking it, the real level of risk is amplified as the patient is still suffering from the original complaint but not taking any medication to bring it under control. If, on the other hand, the same patient underestimates the risk of side-effects, and takes the medication incorrectly or does not tell his or her doctor if a side-effect occurs, the actual level of risk also rises.

### Risk perception by the experts

Risk perception by those who fall into the category of experts is influenced by a different set of factors.

Experts tend to focus their attention on practical measures to deal with risk. Their priorities will include research, monitoring and surveillance. At the same time they do not pay enough attention to the public's need for comprehensible information, nor to their right to participate in decision making [8]. Clearly, the expert's *priorities* differ from those of the public.

Another influential factor is that experts tend to have a conservative attitude towards new data, and do not always revise probabilities in the light of new information [9]. Additionally, it has been observed that they have a tendency to assign higher or lower probabilities to risk data, depending on which they perceive to be more welcome to their superiors [10]. This refers to the *probability assessment* factor.

With regard to *decision-making*, experts tend not to take qualitative attributes of hazardous situations into consideration, instead focusing on what they perceive as objective measures. Thus resources are often allocated to events that are likely to have a large direct impact, while those whose first-order impact is relatively small but could have a far-reaching ripple effect by seriously disrupting social processes are ignored [11].

In the field of *communication* and *participation*, experts consider risk communication more as a means of correcting misconceptions, reducing conflict and educating the public rather than as an end in itself to enable the public to participate in decision making [12]. They discount the public's right to a role in defining levels of acceptable risk and sharing power.

### CONCLUSION

Safety assessment is part of everyone's daily life — it is not the prerogative of experts. This implies that decision making relating to the acceptability of risk must be redefined as a social process of weighing risk against benefit in the context of technical, economic and social factors. No single group has the right to impose its views on any other group; conversely, all groups must be heard and involved in the evaluation process.

In the interests of better risk management, we should identify, catalogue, and understand each other's perceptions and approaches. This will enable us to embark on a dialogue with the other groups involved that makes explicit the various processes of evaluation. Such a dialogue cannot lead to a conflict-free society, but it can, nevertheless, allow the rational resolution of legitimate conflicts [13].

## REFERENCES

[1] Luebbe, H. (1989) Risiko und Lebensbewältigung. In: *Risiko in der Industrie-gesellschaft.* G. Hosemann (ed.), Univ. Bibliothek, Erlangen.

[2] SZ-Magazin (1990) *Süddeutsche Zeitung* Nr. 30, 27 July 1990.

[3] Nowotny, H. & Eisikovic, R. (1990) *Entstehung, Wahrnehmung und Umgang mit Risiken.* Schweizerischer Wissenschaftsrat (eds), Bern.

[4] Bertilsson, M. (1990) The Role of Science and Knowledge in a Risk Society: Comments and Reflections on Beck. In: *Industrial Crisis Quarterly* **4** (2).

[5] Goldstein, B. (1990) The Problem with the Margin of Safety: Toward the Concept of Protection. In: *Risk Analysis* **10** (1) 7–10.

[6] Vlex, C. & Stallen, P. (1980) Rational and Personal Aspects of Risk. In: *Acta Psychologica* **45** 273–300.

[7a] Sandman, P. (1988) Hazard versus Outrage: A Conceptual Frame for Describing Public Perception of Risk. In: *Risk Communication,* Jungfermann *et al.* (eds) Jülich, pp. 163–168.

[7b] Slovic, P., Fischoff, B. & Lichtenstein, S. (1981) Perceived Risk: Psychological Factors and Social Implications. In: *Proc. R. Soc. London* **A376** 17–34.

[8] Vertinsky, I. & Wehrung, D. (1990) Risk perception and drug safety evaluation. *Health and Welfare Canada*, Catalogue no. H42-2/19-19, pp. 1–35.

[9] As [8].

[10] Sanders, F. (1973) Skill in forecasting daily temperature and precipitation: Some experimental results. In: *Bulletin of the American Meteorological Society* **54** 1171–1179.

[11] Vertinsky, I. & Wehrung, D. (1990) *op. cit.*

[12] Kasperson *et al.* (1988) The social amplification of risk: A conceptual framework. In: *Risk Analysis* **8** (2).

[13] Renn, O. (1986) Regulating industrial risks: Science, hazards and public protection. In: CENTED Reprint, no. 56, Clark University, Worcester MA.

# V.3

# Safety assessment in relation to food microbiology

**Anthony Charles Baird-Parker**
Unilever Research, Colworth House, Sharnbrook, Bedfordshire MK44 1LQ, UK

## INTRODUCTION

Assessment of the microbiological safety of food requires, firstly, the identification of those hazards (i.e. those microorganisms and toxins which are of specific concern in the production, distribution and use of a food) and, secondly, the assessment of the probability of conditions arising such that one or more of the identified hazards occurs, and is judged to be an unacceptable risk. Thus, the preferred route of assurance of the microbiological safety of a food is first to carry out a hazard analysis to identify the microbiological hazards and their associated risks throughout the production-use chain, and then to build requirements into all operations to control these — for instance, by specification of product formulation, processing, packaging and distribution conditions, and specific operating requirements (including, where appropriate, instructions for users).

## TECHNIQUES AND PROCEDURES

A variety of hazard analysis techniques and risk assessment procedures can be used to assess safety. These range from relatively unstructured techniques such as 'brain storming' to highly structured ones such as HAZOP (Hazard and Operability Studies). 'Brain storming' has two main disadvantages, namely, expert bias and likely error through failure to understand the often complex interactions of factors in a food operation that can significantly affect the microbiological safety of the food produced.

### Hazard and operability studies (HAZOP)

HAZOP is a multidisciplinary and systematic procedure that can be applied step-wise throughout a food operation by asking a series of structured 'what if' questions concerning potential failures at each stage and considering the probability of occurrence of these, and the consequences of different failure modes [1]. The procedure depends on in-depth information of all aspects of the operation of the plant affecting or potentially affecting the microbiology of the end product (including raw materials, equipment and operating practices). It must be based on up-to-date and detailed knowledge obtained by a technical audit and requires an ability to identify the microbiological implications of normal and abnormal operation of a plant or process. Its main disadvantages are the depth of expertise required, covering areas as broad as engineering, process control, microbiology and hygienic processing. Despite these disadvantages, the technique has much merit as a systematic form of hazard analysis, where hazards and risks can be quantified.

### The Hazard Analysis Critical Control Point (HACCP) system

It is essential that an in-depth hazard analysis is done as a preliminary for setting up *a control system for product safety* such as that based on the use of the HACCP system. Steps in applying HACCP are indicated in Fig. 1.

- Audit food operation to obtain precise data on: raw materials (types, sources and processes used); equipment and facilities (operating parameters, hygienic aspects); operating practices (specifications, monitoring and control procedures); packaging; product storage; intended distribution and use conditions

- Make a flow diagram of operation

- Do a hazard analysis using a multidisciplinary team and applying HAZOP

- Record all hazards according to their severity, and estimate probability of their occurrence

- Identify CCPs, i.e. those places where control can be achieved over a microbiological hazardous raw material, a processing step, an operating practice, etc.

- Specify criteria for control at each CCP

- Identify and document procedures for monitoring that control is achieved (document action to be taken if control tolerances are exceeded)

- Train personnel to carry out their control and monitoring functions

- Validate the HACCP system

Fig. 1 — Procedures to apply HACCP.

### Critical Control Points (CCP)

The identification of Critical Control Points (CCPs), i.e. those places where control over identified hazard(s) can be achieved, is the second stage of the procedure of setting up a HACCP-based control system. Such CCPs may be a raw material or an intermediate product, a piece of equipment, or an operating practice; CCPs may be specific to a particular operation, as also may be the specific control and monitoring procedures chosen to check that they operate effectively. Effective monitoring requires data generated in real time. Thus monitoring will often be based on physical and chemical measurements, such as the temperature of a process or the concentration of salt in a product. Microbiological testing may be appropriate for some CCPs such as raw materials or for verification purposes on end products, but such testing is seldom, if ever, a reliable means of checking that a product is safe [2, 3]. Control systems based on HACCP will give a much higher degree of assurance of safety than those based on microbiological testing. The aim of HACCP is to identify those requirements designed to *prevent* conditions that lead to a microbiologically unacceptable condition occurring, whilst microbiological testing is concerned with looking for defects. Thus microbiological testing is a useful adjunct to control based on HACCP principles but is a poor control procedure *per se*.

## Implementation of HACCP

HACCP should be applied by the operator of a food plant, or a consultant working with an operator, to identify the needs for a specific plant and the specific requirements to control the operation of that plant. Control requirements are usually far too detailed to be widely applied in legislation although some procedures, such as those used for the control of the sterilization of low-acid canned foods, are sufficiently general for CCPs to be documented in legislation. However, the identification of critical control points and the specification and monitoring control requirements in legislation are likely to be the exception rather than the rule. The often-specific nature of the location and control requirements for HACCP means that these must be established in the actual Food Plant. Despite the industry orientation of HACCP some general guidance by the authorities is important. In order for HACCP to be widely applied throughout the Food industry it is important that the responsible Authorities accept its use by industry, and have sufficiently trained personnel to be able to inspect and validate a plant set up on HACCP principles; this requires the thorough training of food inspectors who must have sufficient understanding of the food technology applied, and the microbiological consequence of deviations in control procedures. The EEC is currently sponsoring a FLAIR programme designed to test different methods of applying HACCP within the European food industry and is developing protocols that can be used by different sizes and sectors of the industry. The USA have made progress in applying HACCP in several areas. The USDA Food Inspection Service have set out a detailed strategy for implementing HACCP in meat plants products [4], and the National Marine Fisheries Service has drafted a series of model HACCP plans for fish products and are proceeding to ask industry to file HACCP plans for specific process operations. A recent WHO/EEC sponsored meeting made a series of recommendations to establish procedures for applying HACCP on an international basis [5].

## Risk assessment

Consumer demands and legislative pressures to reduce preservative usage, increased use of less severe processes, and the introduction of novel processing/packaging systems in order to provide better quality and improve convenience may introduce microbiological risks that must be assessed before a product is placed on the market. One traditional means of assessing risk is to carry out a challenge test by inoculating the food with pathogenic microorganisms and then to store the food under conditions that might realistically be expected to occur during its distribution and sale, and storage and use by the purchaser. However, such challenge tests are expensive and slow, and are relatively inflexible as they will only give information on the actual conditions tested. They are being replaced by the use of mathematical models of the kinetics of microbial growth and death which are able accurately to predict the effect of, for instance, storage conditions on the fate of microorganisms in specific food situations [5]. Most models are based on data generated over a wide range of conditions affecting microbial growth or survival such as water activity, pH, temperature and preservatives. Probabilistic and kinetic models of microbial growth have only been widely available recently and many are still under development. However, some predictive models have a very long history of successful use. For instance, the classical means of specifying the heating conditions for the safe thermal processing of low-acid canned foods is based on the use of a model that is able to predict the effects of different time and temperature regimes on the destruction of spores of *Clostridium botulinum*. The UK Government is currently spending more than £1 million per year on a national programme aimed at developing the data, modelling techniques and computer facilities for a national database that can be used for assessing the microbial risks of all types of food. Currently the EC is considering a FLAIR project aimed at developing an EC database and there are strong links with workers in countries as far apart as Australia and USA. The ultimate aim is a database/expert system that can be used for the microbiological safety assessment of foods throughout the world.

### Minimum infective dose (MID)

In any risk assessment, judgements must be made as to the numbers of organisms that are capable of causing infection and disease when consumed in a food. Our knowledge of the minimum infective dose (MID) for many food poisoning microorganisms is poor or very poor. We are aware that the numbers of organisms required to cause infection and disease vary considerably, depending on the food substrate and the response of the individual consuming the food, but have insufficient knowledge of the specific factors that effect the disease state. This area has exercised the minds of clinical microbiologists and epidemiologists for many years. In order to be able to obtain an in-depth opinion as to the current state of knowledge, ILSI (Europe) recently convened a workshop on the concept of MID as applied for safety assessment in food microbiology. As a follow-up to this workshop, Professor Kampelmacher (Bilthoven) and Dr Rowe (PHLS, London) have been invited to review the information concerning MIDs of food-borne illness-causing pathogens and to identify what further information is needed. This will considerably assist the food microbiologist in making an expert assessment of the microbiological safety of foods.

### CONCLUSION

Thus the microbiological safety assessment of a food is becoming established on a more scientific basis, enabling confident predictions to be made about safety and to identify safe routes for the production, manufacture and distribution of all foods.

## REFERENCES

[1] Mayes & Kilsby (1989) *Food Quality and Preference* **1** 53–58

[2] International Commission on Microbiological Specification for Food (1986) *Microorganisms in Foods 2*. (2nd edn) Sampling for Microbiological Analysis: principles and specific applications. University of Toronto Press, Ottawa.

[3] International Commission on Microbiological Specifications for Food (1988) *Microorganisms in Foods, L. Application of the Hazard Analysis Critical Control Point (HACCP) to assure microbiological safety and quality*. Blackwell Scientific Publications, Oxford.

[4] USDA Food Inspection Service (1990) *Hazard Analysis and Critical Control Point (HACCP)*. Implementation–Strategy Paper. FSIS, Washington DC.

[5] Gould, G. W. (1990) *Food Science and Technology Today* **3** (2) 89–92.

[6] World Health Organization (Regional Office for Europe) (1990) Consultation on Food Safety in Europe in the 1990s: The Hazard Analysis Critical Control Point System as the Tool of Choice for Effective Inspection. EUR/ICP/FOS/018(s).

# V.4

# Safety assessment in relation to additives

**Gérard Pascal**
Department of Nutrition, Food and Food Safety, National Institute for
Agricultural Research, 78352 Jouy-en-Josas, France

## INTRODUCTION

Towards the end of 1989, Professor René Truhaut introduced a symposium organized in Brussels by ILSI-EUROPE, dealing with the subject of Acceptable Daily
Intake (ADI). Therefore, I will not elaborate on the concept of ADI and the
modalities of its application to food additives. I refer you to Professor Truhaut's
report [1] and here shall simply quote an extract from Professor Truhaut's speech.
After noting that toxicology had become a multi-field science using many approaches
in very specialized areas such as biochemistry, immunology, molecular biology and
genetics, he declared:

> 'In my opinion, the toxicologist should interpret the results of a given
> approach *within an integrated context*, that is, in conjunction with those
> found by other approaches and taking into account of parameters such as
> physical–chemical properties, structure–activity relationships, conditions of
> use and exposition and toxicokinetic data, especially bioavailability, meta-
> bolic transformation, distribution in body fluids and tissues and rhythm of
> elimination. Such an interpretation of course requires great competence
> based on accumulated experience, *but also intelligence associated with the
> simple common sense needed to arrive at a valid judgement.*'

I agree completely with the conclusion of Professor Truhaut, who considered
that, on this basis, 'the use of the ADI concept has greatly contributed to the
protection of the health of human populations throughout the world, at the same
time facilitating international commerce' [1].

Even if other approaches evaluating the risk presented by the use of food
additives have been proposed, for instance by the Food Safety Council (USA) in
1980 [2] or the 'Red Book' of the FDA (USA) in 1982 [3], which try to quantify the
risk, the ADI concept is still used as a basis for evaluation by committees of experts
such as the JECFA or the SCF of the EC.

This presentation will fall within this framework; its aim is to emphasize, in the light of the author's experience as a member of the SCF and the High Committee on Public Hygiene in France, the main difficulties arising over the last five years in the use of toxicological evaluation of food additives.

I believe that the advantage of such a pragmatic approach lies in the possibility of indicating to industrialists the points to which they should give special attention when composing their files in order to avoid delay in making decisions and to guarantee better consumer safety.

## DIFFICULTIES IN THE CHEMICAL CHARACTERIZATION OF ADDITIVES

There are many examples; however, only three of them are discussed here.

### Mineral hydrocarbons

These products are used directly as additives, or as lubricants during food processing, or as coating agents for packaging materials; their composition varies greatly, partly due to the choice of raw materials. Their processing, particularly in regards to the elimination of unsaturated compounds, has evolved, as well as the composition of the products marketed, especially the minor components.

The toxicological data available on mineral oils, often obtained in conditions which do not correspond to current norms, show wide differences according to the processing used. The chemical composition of the products used in these trials is often not well determined.

Under such conditions, the toxicologist finds it difficult to form a scientific opinion about the risk involved when using these products and that leads him to adopt an extremely prudent attitude.

### Sweeteners (Stevioside)

Stevioside is a sweetening substance found in large amounts in the leaves of *Stevia rebaudiana*, a shrub growing in Paraguay and Brazil. Commercial products widely used in Japan as sweeteners include crude extracts, a purified extract (Stevix) composed of 50–90% of various glucosides (e.g. stevioside, rebauside) and 'Steviosin', a preparation composed of at least 95% of crystallized stevioside.

Toxicological trials have been run on these different preparations. A study on the crude extracts has previously suggested a harmful effect on fertility in rats which could not be reproduced with crystallized stevioside. No adequate study has determined the effects of these different preparations on reproduction.

The only acceptable long-term study was carried out with Stevia extract containing 74.5% of stevioside, 16.3% of rebaudioside A, and 4.4% of another compound; based on this, an ADI of the mixture could be proposed. No evaluation of the long-term effects of extracts of different composition or of stevioside itself has been made because there are no valid data [4].

### Emulsifiers (TOSOM)

Thermally oxidized soya bean oil interacted with the mono and diglycerides of food fatty acids (TOSOM) is an emulsifier which also has anti-spatter properties and can be used in frying-margarine containing proteins.

The composition of this additive is very complex and depends primarily on the conditions in which the thermally oxidized soya bean oil is prepared. A study of its long-term effects and its carcinogenic properties, carried out in satisfactory experimental conditions, were published in December 1987. From the results, an ADI of 25 mg/kg of body weight was proposed for TOSOM [4]. However, the ADI can only be applied to a mixture whose composition closely resembles that used during the recent long-term study; this mixture must be characterized by *precise specifications* ensuring an acceptable level of oxidized fatty acids, polymers and cyclic monomers. No evaluation can be made of mixtures which do not adhere to these specifications.

These three examples lead to the following question: in the case of complex additives which depend on processing methods and which have undergone adequate toxicological testing using a well-characterized mixture, would it not be reasonable, when the results are satisfactory, to give a favourable opinion for inscription of the product on the positive list only if very exact specifications are imposed as well as a method of processing? This procedure is similar to that common in the pharmaceutical field but rare in the food field. However, it is a procedure which would give the consumer maximum safety.

## DIFFICULTIES DUE TO INCOMPLETE KNOWLEDGE OF THE BEHAVIOUR OF FOOD ADDITIVES

In some cases, the consumer not only ingests the additive added to the raw materials but also the products resulting from the reaction of the additive with the food components or with other additives, formed during technological or cooking treatments; the nature and the toxic potentialities of these products are not known. Below are two examples illustrating this problem.

### Flour treatment agents (chlorine)

The treatment of flours with 1500–2500 ppm of chlorine improves the aspect, texture and stability of such products as fruitcake.

99% of the chlorine is bound by the flour. About 20–30% of this chlorine is bound to the flour lipids and about 40–50% is present in the form of chlorides; but about 20% is bound to the carbohydrates and proteins. However, while it has been shown that chlorinated fatty acids are probably not mutagenic, even if they are slightly toxic, the chlorinated amino acids are mutagenic. The case for starch is not known exactly.

Moreover, the toxicology of flour treated with chlorine is not known well enough to make a valid evaluation.

How, in these conditions, without further analytical and toxicological data, can one give a favourable opinion of this treatment?

### Antioxidants (BHT and BHA)

In some countries BHT and BHA are used to prevent oils from oxidation during heating and frying.

Research workers at the Fat Institute in Seville [5] have shown that, when oils are heated, only 63% of the BHT and 30% of the BHA remain in the original state while

26% of the BHT and 60% of the BHA are transformed into products which have not yet been identified and which bind to the fried food. The toxic potentialities of these components are not known and yet they are eaten every day by consumers.

Even if they are not often used together in the same foods, BHA and the nitrites can occur together in a meal and react with each other in the stomach. The mutagenicity of the BHA–nitrite reaction products has just been studied in an *in vitro* system by a German research team [6]. Most of the compounds have been identified; they have no mutagenic potentiality and the work concludes that BHA protects against the formation of nitrosamines in milieus containing nitrites without producing risky components.

Such studies, which give exact information on the behaviour of antioxidants in food and in the organism, are particularly helpful in improving the toxicological evaluation of these additives.

## DIFFICULTIES IN EVALUATING NON-GENOTOXIC CARCINOGENS

Without encroaching on the material of the next chapter, by Professor Kroes on the evaluation of carcinogens, I would like to mention some particularly interesting discussions within the SCF during the latter part of 1990 concerning a colouring agent, erythrosine.

Owing to its chemical structure, this additive acts on the metabolism of the thyroid hormone; the mechanism of this intervention has not been completely elucidated. This hormonal effect is considered to be the cause of the appearance of adenomas and of carcinomas in rat thyroid, at long term, after administration of a strong dose of erythrosine (4% in the diet). This interpretation, which is still open to discussion due to recent genotoxicity studies, has led the SCF as well as the JECFA to attribute an ADI of 0–0.1 mg/kg b.w. to erythrosine; this ADI is based on the dose that had no hormonal effect determined in animals and in humans [4].

The lack of knowledge of the biochemical mechanisms leading to the long-term development of thyroid tumours as a result of thyroid hyperfunction, does not permit, in my opinion, the appplication of an adequate mathematical model for making a quantitative evaluation of the risk of erythrosine intake, although the FDA has taken this step.

The use of mathematical models for the evaluation of the risk due to the ingestion of food additives and contaminants should stimulate a *general reflection* because of the impact on the consumer of this expression of the risks and because of its effects on international regulations.

## DIFFICULTIES DUE TO FAILURE TO RECOGNIZE THE SPECIFIC CHARACTERISTICS OF FOOD TOXICOLOGY AND TO GAPS IN METHODOLOGIES OF TOXICOLOGICAL EVALUATION

### Diet of experimental animals
Interpretation of the results of long-term toxicological trials is often difficult because of a diet which is not well balanced or is responsible for disease affecting all the animals in the experiment.

It has been generally observed that semi-synthetic diets, the composition of which can be perfectly controlled, are not used as much as diets having a composition which is difficult to reproduce in a control study.

In 1984 the SCF could not propose an ADI for the sweetener, neohesperidine dihydrochalcone (NDHC), because its effects on the thyroid could not be explained. A closer look at the results of the USDA works [7] showed that, in the long-term study, all the animals had received a diet subdeficient in iodine; the subdeficiency was increased by the ingestion of NDHC. This was confirmed during a 12-month study using diets containing correct amounts of iodine. In those conditions, no thyroidal hypertrophy was observed.

However, for the SCF to be able to attribute an ADI of 5 mg/kg b.w. to NDHC, it was necessary to make two 90-day complementary studies using a balanced diet; the trials gave satisfactory results [4]. In this way, considerable time was lost.

During the long-term trial on carcinogenesis studying TOSOM already mentioned, a diet of complex composition including barley and oats was used. Although the diet was well-balanced nutritionally, the frequent presence of irritating particles from these diet components was observed between the molars of the animals, in the alveoli and in the nasal and pharynge cavities. At long-term, the permanent irritations caused carcinomas. However these carcinomas, observed with no significant difference in frequency in all the lots of animals, threw a shadow over the results of an experiment which, otherwise, was remarkably well carried out.

## Immuno-toxicology
The evaluation of the allergenic and immunogenic potentialities of several additives has kept us busy for the last few years. Unlike in most important fields of toxicology, where it is usually admitted that 'the dose makes the poison', this statement is not easily applicable here, and we are cruelly lacking a standard methodology.

These problems have often arisen in the case of thickeners, gelifiers (carrageenans) and sweeteners (Sucralose).

International discussions among experts should permit the development of an advanced methodology soon. (For example: a meeting in Paris on 'Practical Aspects of Immuno-toxicology', organized by the International Centre of Toxicology, was held in October 1990.)

## Low-calorie foods
The present fad for low-calorie foods has led to a change in food composition, necessitating the use of intermediate products and thus many additives, such as sweeteners, emulsifiers, thickeners and gelifiers, and bulking agents; these substances are innocuous in some conditions, according to classical toxicological methods but possibly present some problems when large amounts are used (more than 0.5–1% of the food).

The nutritional effects of the use of these additives have still not been adequately determined. The additives may deeply change the physico-chemical conditions in the intestinal medium and affect the microbial ecology in unsuspected ways.

The use of additives which are not very absorbable in the small intestine and are fermentable in the colon, simultaneously with the use of emulsifiers, could modify the digestibility and bioavailability of essential nutrients, particularly vitamins and minerals, but also of contaminants.

These effects must be considered when making up files on these additives which are to be added *into foods*.

*The specificity of food toxicology should be ever present in the minds of promoters of experimental evaluation.*

## SOME PRESENT PROBLEMS

### Sweeteners

In direct relation with what we have been saying, a study of the effects of the use of some additives on the food behaviour of the consumer must be an integral part of the evaluation of their safe use. This is particularly true for sweeteners which have been studied often. I shall mention only one study, published recently in the *American Journal of Clinical Nutrition* [8]. The study included normal-weight subjects receiving, during 3-week experimental periods, either about 1 l of aspartame-sweetened soda, or about 1 l of high-fructose corn-syrup-sweetened soda, or no soda. The results showed that substituting aspartame for fructose syrup avoided weight gain due to the latter and even permitted a slight weight loss in the males, as compared to the experimental period with no soda. The main effect of ingesting aspartame syrup was a drop in the ingestion of simple food sugar.

Such long-term studies more clearly determine the nutritional effects of the use of sweeteners and thus are a better method for the evaluation of the safety of these additives.

### Genetic engineering and biotechnology

A last point to be mentioned is the evaluation of additives produced by biotechnologies using genetic engineering. This concerns enzymes and various additives, but also flavours which are not classified as additives.

Many international committees of experts, including the SCF have been working to define general rules for evaluating these products.

One thing is certain: if it is shown that the structure of the products obtained with these new technologies is identical to that of products made with traditional methods, the method for their examination should not differ from those used up to now.

Most of these compounds are produced by microbes and it is necessary to be sure that the transformed strain is innocuous; this assurance can only be obtained by a *complete* knowledge of the realized genetic construction, the nature of the host, the donor organism of the genetic material introduced and the vector used. This evaluation must be carried out by workers specialized in genetic engineering.

The risk of disseminating genetically modified organisms in the environment should also be evaluated and methodologies defined using a pragmatic approach, case by case.

Finally, the risk, relative to the additive itself, is evaluated essentially on the additive composition, its specifications directly related to the use of the strain, conditions of production and isolation as well as its purification.

This brings us back to an aspect already mentioned at the beginning of this presentation: the safety of an additive is very directly related to its method of production. *Any change* in this area could cause new risks and must be examined for its safety so that new specifications can be set. *Therefore, this should be known by evaluating agencies.*

Today, I think it can be concluded that risk is not necessarily related to the use of genetically-modified organisms, but to additive composition.

## CONCLUSIONS

In the 1950s, efforts began to be made to develop a toxicological methodology for evaluating the risks of the use of food additives; this methodology has been continually improved, partly due to progress in the basic knowledge of xenobiotic action mechanisms. I am not aware of any example showing a real risk for consumer health in the recent past.

However, progress is still needed to increase the safe use of additives; particularly regarding their behaviour in food and during technological and cooking treatments. This remark also applies to pesticide residues and some contaminants.

The efficiency of committees of experts evaluating the innocuity of additives can be improved by:

— a greater effort on the part of industrialists to describe better the production method and specifications of additives;
— a thorough preliminary study, on the part of the promoter, of the toxicological trials and conditions (particularly nutritional) of any experimentation. *There must always be an exchange between the fields of food toxicology and nutrition.* These fields both concern human nutrition. Therefore, the nutritional effects of the use of additives in newly formulated foods should be a point of reflection from the moment the files are drawn up;
— increased cooperation between national and international groups of experts. The help of all these knowledgeable people is needed because of the increasing number of files to be studied. Cooperation to obtain better efficiency is now being planned within the EC where the SCF is using a growing number of basic documents elaborated by national committees.

Experience acquired in the management of the utilization of food additives leads me to think that the problem of food additive safety is not the biggest one. Better efficiency in the treatment of these files will allow experts to devote themselves to other aspects of the relations between food and health; these problems are:

— nutritional imbalance;
— microbiological safety;
— residues and contaminants;
— the risk involved in toxicants naturally present in foods and methods for limiting their levels. This field will certainly be greatly developed.

REFERENCES

[1] Truhaut, R. (1990) The concept of the acceptable daily intake: an historical review. *Food Additives and Contaminants* **8(2)** 152–162.

[2] US Food Safety Council (1980) Proposed System for Food Safety Assessment — Final report of the Scientific Committee, Washington DC.

[3] US Food and Drug Administration (1982) Toxicological Principles for the Safety Assessment of Direct Food Additives and Color Additives Used in Food. Bureau of Foods, Washington, DC.

[4] Reports of the Scientific Committee for Food (1989) Twenty-first series. E.E.C.

[5] Gutierrez Rosales, F., Dobarganes, M. C. & Perez-Camino, M. C. (1986) Accion protectora de los antioxidants BHT y BHA en grasas utilizadas a elevada temperatura. Lecture at The International Symposium on Food Additives in Agro-Food Industry, Madrid, 15–17 Oct. 1986.

[6] Kalus, W. H., Münzner, R. & Filby, W. G. (1990) Isolation and characterization of some products of the BHA–nitrite reaction: examination of their mutagenicity. *Food Additives and Contaminants* 7 (2) 223–233.

[7] Gumbmann, M. R., Gould, D. H., Robbins, D. J. & Booth, A. N. (1978) Toxicity studies of neohesperidin dihydrochalcone. In: 'Sweeteners and Dental Caries', Shaw, J. H. & Roussos, G. G. (eds.); sp. supp. *Feeding, Weight and Obesity Abstracts*, IRL, Washington DC and London, pp. 301–310.

[8] Tordoff, M. G. & Alleva, A. M. (1990) Effect of drinking soda sweetened with aspartame or high-fructose corn syrup on food intake and body weight. *Amer. J. Clin. Nutr.* **51** 963–969.

# V.5

# Safety assessment in relation to carcinogens

**R. Kroes**
National Institute of Public Health and Environmental Protection, PO Box 1,
3720 BA Bilthoven, The Netherlands

## INTRODUCTION

Risk assessment of carcinogenic substances has been a subject of debate during recent decades. During this debate a substantial increase in knowledge was achieved in toxicological phenomena in general and in the process of carcinogenesis in particular. A systemic analysis of the chemical and biological properties and particularly a thorough investigation of the possible mechanism of action of the carcinogen may provide information on which a balanced assessment of risk can be made. In this analysis special attention will be given to the importance of *in vitro* and *in vivo* short-term tests in conjunction with long-term bioassays and other relevant toxicity data. Some results will be shown of evaluations of compounds on the basis of the recommendations concerning the carcinogenicity of chemicals of the Health Council of The Netherlands [18].

## MECHANISM OF ACTION OF CARCINOGENS

Although carcinogens act by a variety of different mechanisms they share a common endpoint, viz. the occurrence of (malignant) tumours. When different mechanisms of action are involved, it seems pertinent to determine whether there should be different approaches to risk estimation for human exposure to carcinogens. Although present knowledge of carcinogenic mechanisms is still limited, we may focus on broadly based concepts in an attempt to classify carcinogens.

One concept which has received world-wide attention is the two-stage model proposed for mouse skin by Berenblum & Shubik [5], which was subsequently shown to be valid for several other tissues [12,14,19,30,31,33,47].

The concept introduced by Miller & Miller [26,27], emphasizing the relevance of metabolic activation of procarcinogens to form electrophilic reactants that interact with DNA which subsequently may lead to irreversibly altered cells is another important aspect for consideration. When we combine these concepts, it may be suggested that the first stage in carcinogenesis, usually called initiation, consists of metabolic activation of the procarcinogen, DNA interaction and 'fixation' of the genetic alteration (see Fig. 1).The next stages, called promotion and progression,

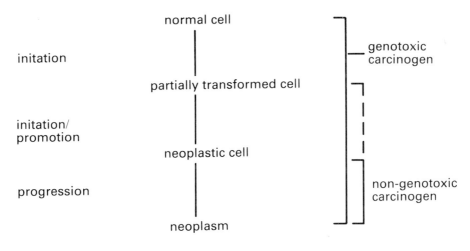

Fig. 1.

represent, respectively, the proliferation of the genetically altered cells to (irreversible) neoplastic cells and the subsequent formation of the tumours through progression of the neoplastic cells [40,43]. Thus in the above-mentioned concept a genotoxic agent may initiate a normal cell to a genetically altered, partially transformed cell. A promotor may complete the conversion of a partially transformed cell to a neoplastic cell and the neoplastic cell will progress to a clinically manifest malignant tumour.

The success of short-term genotoxicity tests in identifying possible carcinogens was based on the rationale that carcinogens may affect DNA. Some carcinogens, however, show negative results in genotoxocity tests, and they may therefore differ from those that do act on DNA. These so-called non-genotoxic carcinogens may thus represent a different class of carcinogens.

## GENOTOXICITY CHARACTERISTICS

Carcinogens having genotoxic properties share, among other things, the following characteristics: they are electrophilic molecules, they alter isolated DNA either by covalent binding or by changing their physical stability, they may induce unscheduled DNA synthesis, they are capable of inducing mutations, they may alter gene expression, and they may cause chromosomal aberrations.

For the detection of genotoxicity, many systems are now available which have been evaluated and validated extensively [2,3,8,15,20,21,37]. The so-called *in vitro* tests contain two components: the microbe, cell or plant in which the genetic change is expressed and a metabolic activation system. Endpoints are, for example, point mutations, deletions, chromosome breaks or transpositions, sister chromatic exchanges and unscheduled DNA repair. In addition to *in vitro* genotoxicity tests, it is of importance to assess genotoxicity *in vivo* [3,4]. It is suggested that the genotoxicity of a chemical can be adequately defined using a combination of the Ames-assay and a test for the assessment or chromosome aberrations *in vitro*. Once a chemical has shown a doubtful response, more *in vitro* tests may be necessary so as to assess the genotoxicity *in vitro* of the chemical. A chemical showing a clear position response, or showing a positive response after more extensive *in vitro* testing, should then always be investigated for genotoxicity *in vivo*, because it has now been unequivocally established that not all *in vitro* genotoxins prove carcinogenic in mammals. *In vivo* short-term tests such as the mouse bone marrow micronucleus assay and, if a negative response is observed, a liver genotoxicity test, may then indicate if carcinogenicity in mammals is likely or not. Substances that are genotoxic but inactive in the two *in vivo* short-term tests mentioned above, may be neither carcinogenic nor mutagenic to germ cells of mammals. It will however be prudent to assess such an absence of carcinogenicity in long-term animal assays.

Thus chemicals with positive *in vitro* genotoxicity results, combined with negative *in vivo* genotoxicity results, with in addition negative carcinogenicity tests in rodents, may be considered to bear negligible risk as to human carcinogenicity and probably mutagenicity in germ cells as well. Genotoxicity tests show significant differences in response for various carcinogens, and their utility for carcinogen prescreening is restricted to the qualitative detection of genotoxic activity, thus predicting rather than defining carcinogenic activity. In evaluating the results of genotoxicity tests, it will become clear that it is often very difficult to reach a conclusion. For example, one or some positive results among many negative results may be noticed. A careful evaluation of such results, in conjunction with many other relevant parameters, as will be discussed later, will be necessary before reaching a conclusion.

## NON-GENOTOXICITY CHARACTERISTICS

Modification of tumorigenic processes is a well-known feature in carcinogen bioassays, but nevertheless an increase in tumour incidence of a 'spontaneously' occurring tumour type is considered to be a carcinogenic effect.

A variety of biological properties are believed to underlie the tumorigenicity of non-genotoxic carcinogens. Examples are non-specific stimulation of cellular proliferation, enzyme stimulation or enzyme inhibition, immune status, hormone balance and nutritional factors [11,15,22,41].

When one suspects a non-genetic mechanism, attempts should be made to obtain more pertinent information concerning the mechanism of action. In a classical two-stage bioassay [6] this non-genotoxic action of a compound can be ascertained when an increased tumour incidence is only found if the compound is given after the administration of an initiating agent, whereas when given before the initiator, no tumours occur. Some *in vitro* short-term tests have been suggested [9,38,44,48] but they will require more thorough validation. These tests are based on the demonstration of a critical property common to most promoting agents that may be detected *in vitro*, such as cell-to-cell communication, or outgrowth of (partially transformed) cells. *In vivo* limited bioassays may provide the possibility of quantitative measurement of non-genotoxic carcinogenicity [43]. Animals are first exposed once or several times to a putative subcarcinogenic dose(s) of a genotoxic carcinogen whereafter they are repeatedly exposed to the test agent to demonstrate supposed tumour enhancement activity by an increased tumour yield. Promoting activity, however, is believed to be tissue specific and thus an appropriate model specifically designed for a target organ is pertinent [35].

## OTHER RELEVANT PARAMETERS TO BE USED IN CARCINOGEN RISK ASSESSMENT

Although similarities in structure to known carcinogens may provide information on possible action or possible pathways in metabolism [26], chemical structure, by itself, is an uncertain predictor of carcinogenic activity. Some attempts have been made to combine structure–activity relationship with short-term test results in the prediction of carcinogenicity of chemicals of unknown carcinogenicity [34].

Biotransformation varies considerably between species, with the age of the test animals, the dose of the carcinogen and the presence of other chemicals (interaction) [42]. Information on metabolism and toxicokinetics is extremely helpful in assessing the mechanism of action and may facilitate the interspecies extrapolation. In appropriate cases, when sufficient information is available the use of physiologically based pharmacokinetic modelling (PBBK) may predict tissue dose parameters for various exposure scenarios in various species thus supporting quantitative risk assessment [1]. Biochemical reactivity, i.e. the degree of adduct formation *in vivo* and *in vitro* , may be explanatory information for potency differences [7]. Physiological effects may influence the carcinogenic response and indeed may be indirectly the reason for a carinogenic response as well. Morphological events such as hyperplasia may be of interest regarding the pathogenesis of the neoplasia involved. In addition, the histogenic site at which a tumour occurs may give additional information: whereas non-genotoxic carcinogens usually affect only one or a few tissues, the occurrence of malignancies at different sites may be a more specific characteristic for genetically active carcinogens [10, 13]. Long-term bioassays usually provide a variety of results which should be considered carefully before the carcinogenicity of a compound can be assessed properly [22]. Important parameters are, for example, latency period, relation to background tumour incidences, histological and biological characteristics of the tumours, dose–response relation and the number of species and strains showing an effect. As mentioned before, limited *in vivo* bioassays may be extremely useful indicators of possible tumour-enhancement properties as well as determinants of initiating properties. Available systems are skin, breast, lung, liver and bladder [19,25,43].

Last but not least, the results of human studies are relevant. Epidemiological studies may provide suggestive evidence for a link between a chemical or an industrial process and carcinogenicity in humans. Such positive findings, in conjunction with the results of bioassays, may be a sound basis for the assessment of the risk in man. Findings which indicate that no relation between a chemical and carcinogenicity can be determined may also be indicative and sometimes be decisive. For example, the extensive epidemiological studies on saccharin have led to the conclusion that this (non-genotoxic) compound may not bear a hazard to man in its current use [28].

## EVALUATION OF RISK OF CARCINOGENS

In long-term bioassays, discrimination between non-genotoxic and genotoxic carcinogens is mostly impossible. Additional testing for (non-) genotoxicity may in certain cases differentiate the two classes. Non-genotoxic or epigenetic carcinogens, or possibly even better carcinogens which exert their carcinogenic action in an indirect way, are believed to show a threshold in their dose–response curve and the existence of the threshold may be used in risk assessment [22,23,24,45]. For such non-genotoxic carcinogens extrapolation is carried out, as is done for non-carcinogens: the determination of a no-effect level, and the use of a safety factor (usually 100 or more).

It should be emphasized, however, that the assumption *a priori* that a non-genotoxic carcinogen is of less concern with respect to human health may be false. Asbestos, not being genotoxic, may be involved in human exposure in quantities sufficient to produce tumours, but exposure to non-genotoxic agents in general has to be considerable to exert an effect.

The presence or absence of genotoxicity is nowadays considered to be a major point in risk assessment to humans. For genotoxic compounds, from a theoretical point of view a threshold cannot be expected, since such substances may be effective even after one single exposure, and may act in a cumulative manner. Thus, in extrapolation to man, the supposed non-existence of a threshold for genotoxic carcinogens will lead to an extrapolation procedure which is conservative. In practice, genotoxic compounds will as much as possible be eliminated from the environment, and new substances will not be introduced based on their genotoxic properties. If, however, such substances cannot be eliminated, conservative extrapolation procedures should be used to assess the risk for man.

Many mathematical models have been suggested for extrapolation of genotoxic carcinogens [40]. When such models are used, the notion of absolute safety is impossible to achieve and should be replaced by the concept of virtual safety. Whereas mathematical models may be appropriate to estimate low-dose incidences in the animal species under investigation, they do not address the problem of conversion of such animal data to man. The calculation of virtual safe doses should be done at risk levels in the range of practical and social concern. The model chosen may depend on the experimental data. For regulatory purposes in The Netherlands, a linear extrapolation model is used, using the lowest effective dose. Only when the experimental data permit another approach, are other models used.

A major point of concern, however, is the fact that a clear distinction between direct-acting genotoxic and indirect-acting, most often non-genotoxic carcinogens, is often not possible. Moreover, genotoxic compounds may also exert their carcinogenic action merely by a non-genotoxic mechanism. Therefore it is proposed to evaluate carcinogens on an individual basis, using all available published and non-published information.

In The Netherlands a special committee under the guidance of the Health Council has been created for this specific purpose. This committee considers carcinogenic compounds for which questions arise, and they are assisted in this by *ad hoc* experts. The National Institute of Public Health and Environmental Protection usually provides the review of literature on the compounds and proposes a classification and risk assessment which is then reviewed by the expert committee.

## RISK ASSESSMENT TO MAN

If the risk of carcinogens is to be considered, one of the first aspects to investigate is the possible genotoxicity. Additional information on structure, biotransformation, toxicokinetics functional and morphological effects, and the results of (limited) bioassays may enhance or decrease the likelihood of genotoxic activity in man.

Genotoxic activity of a carcinogen usually leads to a conservative approach in risk assessment for man. If, however, evidence is available that proven genotoxicity in experimental animals is not likely to occur in man, owing to, for example, differences in metabolism, one may decide to use another appropriate approach in risk assessment. Absence of genotoxic activity for carcinogens warrants a conventional approach, since a threshold phenomenon can be expected.

Concerning genotoxic carcinogens the best approach is the banning of the substance from the environment. If, however, elimination seems not feasible, linear extrapolation using the lowest effective dose is suggested when assessing risk to man. The extrapolated animal data to very low dosages with negligible risk (usually one cancer in a million in a lifetime) should then be converted to man. Assuming that man is not more sensitive than the experimental animal, a direct conversion is usually accepted for a number of reasons:

(1) The linear extrapolation is most conservative. It calculates the theoretical maximum incidence, and it is very likely that the actual incidence will be (much) lower.
(2) Metabolic activation in man usually takes place at a lower rate as compared to experimental animals and is inversely related to lifespan and body weight [18,32,36].
(3) DNA repair processes seem positively related to lifespan [16,39].
(4) The stability of genetic information of organisms increases with body size and brain weight [17].
(5) Sensitivity of humans to known carcinogens is about equal to or less than that of experimental animals [18,29,40,46].

The discrimination between non-genotoxic and genotoxic action is usually not as clear as one would like it to be. It is therefore important that an expert-panel considers all available data to make up a final judgement. In certain cases, incidental positive findings in genotoxicity tests can be ascribed to (high) concentration phenomena and the like. In addition, some carcinogens having genotoxic properties may act as a non-genotoxic carcinogen in carcinogenesis. It seems likely that a range of carcinogens, starting with pure initiatiors on the one end and ending with pure promotors on the other end, can be identified. Thorough consideration of the data by experts in the field may provide the tools required to reach a well-balanced evaluation and to assess adequately the carcinogenic risk to man. In The Netherlands the expert committee on the evaluation of carcinogens has evaluated a number of substances in the way described above (see Fig. 2). These evaluations have resulted

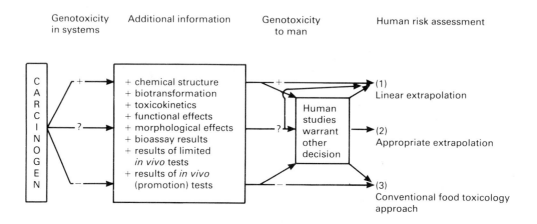

Fig. 2. — Risk assessment procedure for carcinogens in The Netherlands (after Kroes [24]).

in proposals to consider some of them as threshold carcinogens; these are: saccharin, NTA, TCDD, trichloroethene, tetrachloroethene, trichloromethane (chloroform), carbon tetrachloride, methylene chloride, cadmium, lead, formaldehyde and styrene. For other compounds it has been proposed to consider them as non-threshold carcinogens with subsequent linear extrapolation in risk assessment to man. Examples are: 2-nitropropane, aflatoxin, vinylchloride, acrylonitril, ethylene oxide, dichloroethene, epichlorohydrin, nitrosamines, chromium VI and polycyclic aromatic hydrocarbons.

Risk assessment in man, especially for carcinogens, has been and will be a controversial issue. It has to be realized, nevertheless, that man may be exposed to carcinogens every day. Some of these carcinogens can be banned and eliminated; others, however cannot be eliminated at all. Moreover, there may exist valid social or economic reasons to accept exposure to carcinogens. It is therefore of vital importance that in assessing risks, one discriminates between real risks (i.e. vinylchloride or aflatoxins) and relatively minor or even non-risks (as for example saccharin or NTA). In assessing risks to man, such quantified risks should be compared to other risks to man in order to make a proper judgement concerning their significance. Moreover, comparison of risk will, in particular, make the general public aware that not all carcinogens are alike and neither are their risks.

## ACKNOWLEDGEMENT

The author acknowledges the excellent administrative assistance of Miss A. M. B. Marks and Mrs C. Exel in preparing this manuscript.

## REFERENCES

[1] Anderson, M. E. (1989) Tissue dosimetry, physiologically-based pharmaco-kinetic modelling, and cancer risk assessment. *Cell Biology and Toxicology* **5** 405–415.

[2] Ashby, J., de Serres, F. J., Draper, M., Ishidate, M., Margolin, B. H., Matter, B. & Shelby, M. D. (1985) Overview and conclusions of the IPCS collaborative study on *in vitro* assay systems. In: J. Ashby, F. J. de Serres, M. Draper, M. Ishidate, B. H. Margolin, B. E. Matter & M. D. Shelby (eds) *Evaluation of short-term tests for carcinogens*, vol. 5. Elsevier Science Publishers, Amsterdam, Oxford, New York, pp. 118–174.

[3] Ashby, J. (1986) Discussion forum. The prospects for a simplified and harmonized approach to the detection of possible human carcinogens and mutagens. *Mutagenesis* **1** 3–16.

[4] Ashby, J. (1990). Genotoxicity testing: to what extent can it recognize mutagens and carcinogens? In: D. B. Clayson, I. C. Munro, P. Shubik & S. A. Swenberg (eds) *Progress in Predictive Toxicology*. Elsevier Science Publishers, Amsterdam, pp. 185–205.

[5] Berenblum, I. & Shubik, P. (1947) The role of croton oil applications, associated with a single painting of a carcinogen, in tumour induction of the mouse's skin. *Br. J. Cancer* **1** 379–383.

[6] Berenblum, I. (1982) Sequential aspects of chemical carcinogenesis: skin. In: F. F. Becker (ed.) *Cancer: a comprehensive treatise.* vol. 1, 2nd edn. *Etiology: Chemical and physical carcinogenesis,* Plenum Press, New York, pp. 451–484.

[7] Booth, S. C., Bosenber, H., Garner, R. C., Hertzog, P. J. & Morpoth, K. (1981) The activation of aflatoxin $B_1$ in liver slices and in bacterial mutagenicity assays using livers from different species including man. *Carcinogenesis* **2** 1063–1068.

[8] Bridges, B. A. (1988) Genetic toxicology at the crossroad — a personal view on the deployment of short-term tests for predicting carcinogenicity. *Mutation Research* **205** 25–31.

[9] Brookes, P. (1981) Critical assessment of the value of *in vitro* cell transformation for predicting *in vivo* carcinogenicity of chemicals. *Mutat. Res.* **86** 233–242.

[10] Clayson, D. B. (1975) the chemical induction of cancer. In: E. J. Ambrose & R. J. C. Roe (eds) *Biology of cancer.* Ellis Horwood, Chichester, pp. 163–179.

[11] Clayson, D. B. (1989) Can a mechanistic rationale be provided for non-genotoxic carcinogens identified in rodent bioassays? *Mut. Res.* **221** 53–67.

[12] Clifton, K. H., Sridharin, B. N. (1975) Endocrine factor and tumor growth. In: F. F. Becker (ed.) *Cancer: a comprehensive treatise,* vol. 3. Plenum Press, New York, pp. 249–286.

[13] Doull, J., Klaassen, C. D., Amdur, M. O. (1980) *Cassarett and Doull's toxicology, the basic science of poisons,* 2nd edn. Macmillan, New York.

[14] Farber, E. & Cameron, R. (1980) The sequential analysis of cancer development. *Adv. Cancer Res.* **31** 125–226.

[15] Grice, H. C. (1984) *Current issues in toxicology. Extrapolation of toxicity data.* Springer, Berlin, Heidelberg, New York, Tokyo.

[16] Hart, R. W., Setlow, R. B. (1974) Correlation between deoxyribonucleic acid excision-repair and life-span in a number of mammalian species. *Proc. Natl. Acad. Sci. USA* **71** 2169–2173.

[17] Hart, R. W., Turturro, A. (1983) Species longevity as an indicator for extrapolation for toxicity data among placental mammals. *J. Am. Col. Toxicol.* **2** 235–243.

[18] Health Council of the Netherlands (1980) *The evaluation of the carcinogenicity of chemical substances.* Government Office, The Hague.

[19] Hicks, R. M., Chowaniec, J., Wakefield, J. St. J. (1978) Experimental induction of bladder tumors by a two-stage system. In: T. J. Slaga, R. K. Boutwell & A. Sivak (eds) *Mechanism of tumor promotion and cocarcinogenisis*, vol 2. Raven Press, New York, pp. 475–490.

[20] ICPEMC (1988) Testing for mutagens and carcinogens: the role of short-term genotoxicity assays. Publication no. 16. *Mutation Research* **205** 3–12.

[21] IPCS (1988) *Evaluation of short-term tests for carcinogens., Report of the International Programme on Chemical Safety's Collaborative Study on* in vivo *assay.* J. Ashby *et al.*, (eds) Volume 1. 1.1–1.435. Volume 2. 2.1–2.376. Cambridge University Press on behalf of WHO. Cambridge, england.

[22] Kroes, R. (1979) Animal data, interpretation and consequences. In: P. Emmelot & E. Kriek (eds) *Environmental carcinogenesis.* Elsevier/North-Holland, Biomedical Press, Amsterdam, pp. 297–302.

[23] Kroes, R. (1983) Short-term tests in the framework of carcinogen risk assessment to man. In: G. M. Williams, V. C. Dunkel & V. A. Ray (eds) *Cellular systems for toxicity testing, Ann. NY Acad. Sci.*, vol. 407. New York Acad. Sci., New York, pp. 398–408.

[24] Kroes, R. (1987) Contribution of toxicology towards risk assessment of carcinogens. *Arch. Toxicol.* **60** 224–228.

[25] Mazue, G., Gowy, D., Remanda, B. & Garbay, J. M. (1983) Limited *in vivo* bioassays. In: G. M. Williams, V. C. Dunkel & V. A. Ray (eds). *Cellular systems for toxicity testing, Ann. NY Acad. Sci.*, vol. 407. New York Acad. Sci., New York, pp. 374–386.

[26] Miller, E. C. & Miller, J. A. (1976) The metabolism of chemical carcinogens to reactive electrophiles and their possible mechanism of action in carcinogenesis. In: C. E. Searle (ed.) *Chemical carcinogens.* Am. Chem. Soc., Washington DC, pp. 737–762.

[27] Miller, E. C. & Miller, J. A. (1986) Electropholic sulfuric acid ester metabolites as ultimate carcinogens. In: Kocsis *et al.* (eds) *Biological Reactive Intermediates III: Mechanisms of action in animals, models and human disease.* pp. 583–596. Plenum Press, New York.

[28] Morgan, R. W., Wong, O. (1985) A review of epidemiological studies on artificial sweeteners and bladder cancer. *Food Chem. Toxicol.* **23** 529–533.

[29] National Academy of Sciences (1975) Consultative panel on health hazards of chemical presticides. *Pest Control, vol. 1: An assessment of present and alternative technologies.* Washington DC.

[30] Peraino, C., Fry, R. J. M., Stoffeldt, E. & Christopher, J. P. (1975) Comparative enhancing effects of phenobarbital, amobarbitol, diphenylhydantoin and dichlorodiphenyltrichloroethane on 2-acetylaminofluorine-induced hepatic tumorigenesis in the rat. *Cancer Res.* **35** 2884–2890.

[31] Pereira, M. A. & Stoner, G. D. (1985) Comparison of rat liver foci assay and strain. A mouse lung tumor assay to detect carcinogens. A review. *Fund Appl. Toxicol.* **5** 688–699.

[32] Ramsey, J. C. & Gehring, P. J. (1980) Application of pharmacokinetic principles in practice. *Fed. Proc.* **39** 60–65.

[33] Reddy, B. W., Weisburger, J. H. & Wynder, E. L. (1978) Colon cancer: bile salts as tumor promotors. In: R. K. Slaga, R. K. Boutwell & A. Sivak, (eds) *Mechanisms of tumor promotion and cocarcinogenesis,* vol. 2. Raven Press, New York, pp. 453–464.

[34] Rosenkranz, H. S., Klopman, G.,, Chankong, V., Pet-Edwards, J. & Haimes, Y. Y. (1984) Prediction of environmental carcinogens: a strategy for the mid-1980s. *Environ. Mutagen.* **6** 231–258.

[35] Shubik, P. (1984) Progression and promotion. *JNCI* **73** 1005–1011.

[36] Sokal, J. A. (1982) Extrapolation of experimental data to humans. In: *Evaluation and risk assessment of chemicals.* WHO, Copenhagen, pp. 139–149.

[37] Tennant, R. W. *et al.* (1987) Prediction of chemical carcinogenicity in rodents from *in vitro* genetic toxicity assay. *Science* **236** 933–941.

[38] Trosko, J. E., Yotti, L. P., Warren, S. T., Tsushimoto, S. G. & Chang, G. (1982) Inhibition of cell–cell communication by tumor promotors. *Carcinogen Compr. Surv.* **7** 565–585.

[39] Turturro, A. & Hart, R. W. (1984) DNA repair mechanisms in aging. In: D. G. Sciapelli & G. Migaki (eds) *Comparative biology of major age-related diseases: current status and research frontier.* A. R. Liss, New York, pp. 19–45.

[40] US Office of Science and Technology Policy (1985) Chemical carcinogens: A review of the Science and its associated principles February 1985. *Fed. Reg.* Part. II March 14: 10371–10442.

[41] Weisburger, J. H., Williams, G. M. (1980) Chemical carcinogens. In: J. Doull, C. D. Klaassen & M. O. Amdur (eds) *Toxicology: The basic science of poisons,* 2nd edn. Macmillan Press, New York, pp. 84–138.

[42] Weisburger, J. H. & Williams, G. M. (1982) Metabolism of chemical carcinogens. In: F. F. Becker (ed.) *Cancer: a comprehensive treatise,* 2nd edn. Plenum Press, New York, pp. 241–333.

[43] Weisburger, J. H. & Williams, G. M. (1989) Types and amounts of carcinogens as potential human cancer hazards. *Cell Biol. and Tox.* **5** (4) 377–392.

[44] Williams, G. M. (1981) Liver carcinogenesis: the role for some chemicals of an epigenetic mechanism of liver-tumor promotion involving modification of the cell membrane. *Food Cosmet. Toxicol.* **19** 577–583.

[45] Williams, G. M. & Weisburger, J. H. (1981) Systematic carcinogen testing

through the decision point approach. *Ann. Rev. Pharmacol. Toxicol.* **21** 393–416.

[46] Williams, G. M., Reiss, R. and Weisburger, J. H. (1985) A comparison of the animal and human carcinogenicity of environmental, occupational and therapeutic chemicals. In: W. G. Flamm and R. J. Lorentze (eds) *Adv. in Modern Environmental Toxicology, Vol. VII. Mechanisms and Toxicity of Chemical Carcinogens and Mutagens.* Princeton Sci. Publ., Princeton NJ, pp. 207–248.

[47] Witschi, H., Lock, S. (1978) Butylated hydroxytoluene: a possible promotor of adenoma formation in mouse lung. In: T. J. Slaga, R. K. Boutwell & A. Sivak, (eds) *Mechanisms of tumor promotion and cocarcinogenesis,* vol 2. Raven Press, New York, pp. 465–474.

[48] Yamasaki, H. (1984) Modulation of cell differentiation by tumour promotors. In: T. J. Slaga (ed.), *Mechanisms in tumour promotion,* Vol. IV, CRC Press, Boca Raton FL, pp. 1–26.

## Part VI
# Poster session
# (36 contributions)

## VI.1 THE USE OF AN AUTOMATED TURBIDIMETRIC ASSESSMENT OF THE DETECTION TIME AS AN ALTERNATIVE TO THE CLASSICAL AEROBIC COLONY COUNT

**P. van Netten** and **B. van Lith**
The Inspectorate for Health Protection, 's-Hertogenbosch, The Netherlands
**A. Dolfing**
The Inspectorate for Health Protection, Groningen, The Netherlands

The Dutch Food Act ('*warenwet*') includes microbiological standards for the classical aerobic colonization of foods. The classical aerobic colony count determination is obligatory, but it is laborious and expensive. Moreover, the aerobic colony count standards are only exceeded by up to approximately 40 per cent of the routinely examined samples. Therefore, a splitting of the routine samples, before the obligatory classical aerobic colony count assessment, into samples suspected of a too high aerobic colony count and unsuspected samples can be cost-effective.

The use of the Cobas Bact turbidimetric centrifugal spectrofotometer has been tested in estimating the aerobic colonization of minced meat, chinese meals, crumbed snacks and ice-cream in terms of aerobic colony count. Pure culture studies with a selection of Gram-positive and Gram-negative bacteria had revealed linear regression coefficients of 0.91–0.97. Furthermore, colony counts equivalent to the standards in the Dutch Food Act, that is, exceeding $10^{5-6}$ cfu/g or ml, could be detected within 6–8 h. Strict aerobes, that is *Pseudomonas* spp. and microccocci differed by up to 1.5 h in their detection time from that of facultative Enterobacteriaceae and Staphylococcci, respectively.

Linear regression correlation coefficients between the aerobic colony counts and the automated turbidimetric assessed detection times varied from 0.68 to 0.75 for ice-cream and from 0.80 to 0.90 for the other retail foods tested.

The automated centrifugal turbidimetric apparatus allowed determination in less than 9 h, that is one-third of the time needed for a conventional aerobic colony count, whether retail foods were in compliance with the aerobic colony count standard in the Dutch Food Act. The percentage of food samples for which an obligatory classical assessment of the aerobic colony count was necessary reduced from 100 per cent to 20 per cent for crumbed snacks and to 34 per cent for chinese meals. However, the automated turbidimetric assessment of the detection time was less effective in the selection of minced meat, ice-cream and whipped cream samples with too high aerobic colony counts, resulting in a percentage of samples having to be re-examined by the classical colony count, the percentage being 57–62.

## VI.2   EVALUATION OF THE MICROBIOLOGICAL QUALITY OF MINCED MEAT

**Gudrun Depourcq** and **Lieve Van Poucke**
Laboratory for Pharmaceutical Microbiology and Hygiene, University of Ghent, Harelbekestraat 72, B-9000 Ghent, Belgium

Raw meat is always infected with a certain amount of microorganisms. During production, handling and/or conservation, the initial infection grade can change quantitatively as well as qualitatively. The meat becomes rotten or food-poisoning due to pathogenic bacteria can occur.

Especially in industrialized countries, modern lifestyle has changed eating habits.

Changes in handling and feeding of animals, the mingling of animals before slaughter and cross-contamination of carcasses on production lines, fresh-marketing practices that prolong the period before products are sold, and lower salt, nitrite and acid usage all contribute to the problem of microbiological contamination.

Fifty-two samples of minced meat were examined for their bacteriological quality.

The examination consisted of determination of the total amount of germs, the search for Enterobacteriaceae, *Escherichia coli* and faecal *Streptococcci* as indices for faecal contamination, *Salmonella, Staphylococcus aureus* and *Clostridium perfringens* as pathogenic bacteria, and *Staphylococci* and yeasts as indications for a bad or a too long conservation.

The norms for the parameters used are those for frozen minced meat, established by Belgian law: $10^6$ germs per gram for the total amount of bacteria; $10^3$ per gram for Enterobacteriaceae; $<100$ per gram for *Escherichia coli*; $10^3$ per gram for faecal *Streptococci*; 0 in 25 gram for *Salmonella*; $<100$ per gram for *Staphylococcus aureus*; $<30$ per gram for *Clostridium perfringens*.

The following results were obtained. In 25/51 of the cases the total amount of bacteria was too high. Seventeen species of Enterobacteriaceae were isolated, in 43/52 of the samples in amounts which exceeded the norm. But those Enterobacteriaceae are not dangerous for the consumer. Faecal *Streptococci* were found in 23 cases and *Escherichia coli* in 6, of which 4 had norm-exceeding quantities. This is a clear indication for faecal contamination and thus a poor hygiene. *Salmonella* was found in 2 samples. *Clostridium* and *Staphylococcus aureus* in 4, resp. 4, of which 1, resp. 4, was in amounts which exceeded the norm. *Staphylococci* were detected in all the specimens, except 3, yeasts in 36 and moulds in 7. This indicates a too long or bad conservation.

The screening leads to the conclusion that 13 samples were of bad quality, 32 of lower but acceptable quality, and only 7 out of 52 of good quality.

Improvements in the production equipment and better hygienic procedures to reduce contamination of meat with faecal materials, the principal source of contaminating pathogens, are recommended.

## VI.3   ONE-DAY DETECTION OF *SALMONELLA* IN FOODS

**J. De Smedt**
Jacobs Suchard, Montezumalaan 1, B-2410 Herentals, Belgium

For the food industry, timely identification of contaminated products is highly important in the microbiological safety assurance of the production process and its environment. The introduction of motility enrichment on Modified Semisolid Rappaport–Vassiliadis (MSRV) medium (De Smedt & Bolderdijk, 1987) is a sensitive diagnostic tool for rapid *Salmonella* detection from foods, based on the acceptance criteria of sensitivity, cost-efficiency and simplicity. The procedure consists of pre-enrichment followed by motility enrichment on MSRV in petri dishes. The efficiency of this medium is based on the ability of *Salmonellae* to migrate through the highly selective medium, while competitive bacteria are inhibited. Serological tests can be performed on migrated cultures, so that a confirmed test result is obtained within 48 h after the beginning of the test.

By reducing the pre-enrichment time, even one-day detection is possible. In this procedure, M-broth is used as the pre-enrichment medium, while the pre-enrichment culture is incubated for 8 h by stirring incubation. For the motility enrichment a slightly modified MSRV medium (called MSRVa) is used. In this MSRVa medium, the pH is increased from 5.2 to 6.4, which enhances migration so that migration is obtained after only 16 h of incubation. In this way, test results are obtained after 24 h. The procedure has been tested with naturally and artificially contaminated food samples and has shown that, depending on the type of sample, 90–100 per cent of positive samples can be detected. This gives the possibility to take immediate action, after only one day of testing, to remove sources of contamination.

### Reference
De Smedt, J. M. & R. Bolderdijk (1987) Dynamics of *Salmonella* isolation with Modified Semisolid Rappaport–Vassiliadis medium. *J. Food Prot.* **50** 658–661.

## VI.4   DETECTION OF CAMPYLOBACTER IN PROCESSED POULTRY

**L. De Mey, W. De Greyt, H. Fierens** and **A. Huyghebaert**
Laboratory of Food Technology, Chemistry and Microbiology, Faculty of Agricultural Sciences, University of Ghent, Coupure links, 653 B-9000 Ghent, Belgium

The presence of Campylobacter in processed poultry was investigated using different procedures. Direct isolation was compared with enrichment techniques.

For direct isolation a certain amount (5–20 g) was digested during 45 s with dilution fluid to obtain the $10^{-1}$ dilution. Brucella broth was chosen for the dilution as the solubility of oxygen in this medium is much lower (7 versus 13 ppm) than in ordinary Ringer solution or other fluids. A number of different isolation media were tested: Butzler, Skirrow, Blaser-Wang, Preston. All these media use an agar base with addition of different supplements. The agar base was either Columbia agar or Campylobacter agar. Supplements consisted of growth factors and redox reducers such as succinate, pyruvate, sulphate, sulphite, cysteine and several mixtures of antibiotics such as vancomycine, trimetoprim, polymyxin, bacitracin, colistine, actidione, novobiocin, cephazoline; sterile defibrinated horse blood was also added. Isolation was performed by streaking 0.1 ml on the surface of the different media. Incubation was at 42°C in an anaerobic jar.

For the enrichment method two basic media were used: either Brucella broth supplemented with growth factors alone or the same medium supplemented with growth factors and several antibiotics (the Doyle & Roman medium). Incubation was at 42°C in an anaerobic jar with subsequent isolation on the aforementioned media.

The most selective agar media proved to be Butzler and Skirrow with no distinction between direct isolation or enrichment.

The best enrichment medium was modified Doyle & Roman medium (pyruvate added).

Of the total number of samples found positive after enrichment, only 1/3 could be isolated by the direct method. This might indicate that the number of Campylobacter cells present was too low to be detected, or that the recovery from the poultry samples was too low.

To test this hypothesis, several sterile chicken fillets were inoculated with different concentrations of Campylobacter cells. Immediately after inoculation the total number of Campylobacter present was determined by direct isolation, the same procedure being repeated after 24 h in cool storage.

Direct isolation after inoculation gave a recovery of approximately 50 per cent of the total number inoculated, which means that about 200 Campylobacter per g can be detected by direct isolation. This number is probably seldom present in processed poultry (as later research proved). The same samples analysed after 24 h in cool storage gave a supplementary reduction of about 25 per cent in total number, the overall recovery being thus less than 25 per cent.

In conclusion it can be said that an enrichment procedure seems to be necessary for efficient and reliable isolation of Campylobacter from processed poultry.

## VI.5   THE OCCURRENCE AND FATE OF *LISTERIA MONOCYTOGENES* IN MEAT AND MEAT PRODUCTS: A LEGAL CONSIDERATION

**P. van Netten**
The Inspectorate for Health Protection 's-Hertogenbosch, The Netherlands
**E. de Boer**
The Inspectorate for Health Protection Zutphen, The Netherlands
**J. T. Jansen**
The Inspectorate for Health Protection Rijswijk, The Netherlands

Outbreaks of human listeriosis in the last decade and the foods implicated support the concept of zoonotic transmission of listeriosis. Presently, direct evidence for meat as a vector in human listeriosis is lacking, but there is some circumstantial evidence that suggests a role for meat and meat products.

For public health protection a standard has been suggested that enforces, even up to sell-by date, the absence of *L. monocytogenes* in 0.01 g food that is, less than 100 cfu of *L. monocytogenes* per g (Jansen (1990) *Ware(n)-chemicus* **20** 102–118). A major advantage of the proposed standard is that the assessment of *L. monocytogens* in meat and meat products is simplified to a rapid, simple and reliable counting procedure, as for instance the PALCAM blood agar overlay procedure developed by van Netten *et al.* (*Letters in Applied Microbiol.* In press.)

The consequences of the proposed standard for meat and meat products may be evaluated as follows.

To enforce the absence of *L. monocytogenes* in 25 g samples of fresh meat is unrealistic. A small survey demonstrated that *L. monocytogenes* prevails in 11 per cent of heat-processed meat, in 40 per cent of fermented sausages and in 55–70 per cent of raw meat and poultry. However, a standard for the absence of *L. monocytogenes* in 0.01 g meat accepts the ineluctable contamination of raw meat with low numbers of *L. monocytogenes*, but it prohibits fresh meat in which *L. monocytogenes* has proliferated due to refrigeration abuse and too long chilled storage. A nation-wide survey revealed that up to 4 per cent of fresh meat samples examined was colonized with more than 100 cfu of *L. monocytogenes* per g and thus violated the proposed standard.

An increase of *L. monocytogenes* in heat-processed meat does not take place in the presence of $10^{6-8}$ lactic acid bacteria. A 100-fold increase in *L. monocytogenes* count takes place in sterile heat-processed pâté of pH 6.0 within 12 days of chilled storage at 3°C, within 8 days at 7°C and within 2.5 days at 12.5°C. Therefore, the standard prohibits heat-processed meat in which *L. monocytogenes*, recontaminating after heat processing, has proliferated during distribution and retail storage. A small survey revealed a violation of the standard by 1 per cent of 83 tested samples of pâté from retail stores.

The *L. monocytogenes* counted in fermented sausages remains more or less stable during up to 25 days of chilled storage (2 and 7°C), refrigeration abuse (12.5°C) and declines during storage at 22°C. Hence, for fermented meat the standard implies the interdiction of the use of chilled stored raw meat colonized with high cfu of *L. monocytogenes* and *Pseudomonas* spp. In a survey, 3 per cent of 268 tested fermented sausages counted more than 100 cfu of *L. monocytogenes* per g.

It may, therefore, be concluded that the absence of *L. monocytogenes* in 0.01 g fresh meat and meat products is a reasonable standard.

## VI.6 PSYCHROTROPHIC STRAINS OF *BACILLUS CEREUS* PRODUCING ENTEROTOXIN

**P. van Netten, P. van Hoesel** and **A. van de Moosdijk**
The Inspectorate for Health Protection 's-Hertogenbosch, The Netherlands

In investigation of three outbreaks of *Bacillus cereus* food poisoning in Spain and The Netherlands, the causative strains demonstrated growth within a temperature range of 4–37°C. Such psychrotrophic types were found to occur in dairy products (including ca. 25 per cent of 35 samples of pasteurized milk), in some mousses and cooked/chill meals.

Enterotoxin formation was unaffected by food attributes as long as $a_w$ was at least 0.95 and pH values were no lower than 5.8. Enterotoxin production was produced after 24 days at 4°C, about 12 days at 7°C and within 48 hours at 17°C — the temperature used to mimic slight temperature abuse of cooked/chill meals and pasteurized milk. Detectable enterotoxin concentrations coincided with colonization levels above $10^{6.6}$/g.

In experiments wherein temperature abuse of cooked/chill meals and pasteurized milk was mimicked, keeping commodities for some hours at 17°C or for 1 day at 12.5°C appeared to shorten, by 2–6 days, the time required for formation of enterotoxin during subsequent storage at 7°C.

The results indicate that, unless cooked/chill meals are stored at below 4°C throughout, they present a risk of *B. cereus* toxication, in addition to that of transmitting listeriosis, versiniosis and botulism. Furthermore, chilled storage (7°C) of pasteurized milk may lead to toxin, detectable with Oxoid's BCET-RPLA latex agglutination test-kit up to 12 days before overt spoilage.

## VI.7  HPLC DETERMINATION OF AFLATOXINS IN FOOD BY POST-COLUMN DERIVATIZATION WITH ELECTROCHEMICALLY GENERATED BROMINE

**M. C. Spanjer, J. M. Scholten** and **A. E. Strooper**
Governmental Food and Commodities Inspection Service, Burgpoelwaard 6, 1824 DW Alkmaar, The Netherlands

According to the Dutch Food Act aflatoxins have to be determined by means of two-dimensional TLC. This is, however, a rather time-consuming and laborious method. Apart from this, it can be difficult to quantify the spots that are obtained. Another problem is the lack of difference between several kinds of aflatoxins. For these reasons we have tested the applicability of the method Kok *et al.* [1] developed for determination of aflatoxins in cattle feed. They enhanced the native fluorescence of aflatoxin by derivatization with on-line, electrochemically generated bromine. This is post-column produced from bromide which is dissolved in the mobile phase. Separation between aflatoxins B1, B2, G1 and G2 is achieved within 15 minutes in a reversed-phase system. Since the tolerance for human food is lower than for cattle feed, we investigated its usefulness for food control purposes. Optimization of the extraction has also been sought [2]. Originally, cleanup was carried out with silica. This is compared with recent experiments with immunoaffinity cleanup [3]. The results of this effort will be presented for peanut, peanut butter, pistachio nut, fig and sauce samples. Because of the inhomogeneous distribution of contaminated nuts some information about sampling and sampling preparation is given.

**References**
[1] Kok, W. T., van Neer, T. C. H., Traag, W. A. & Tuinstra, L. G. M. T. (1986) *J. of. Chromatography* **367** 231.
[2] Whitaker, T. B., Dickens, J. W. & Giesbrecht, F. G. (1986) *J.A.O.A.C.* **69** 508.
[3] Shepherd, M. J., Mortimer, D. N. & Gilbert, J. (1987) *J. Assoc. Publ. Analysts* **25** 129.

## VI.8 PRODUCTION OF ANTIBODIES TO HAZELNUT PROTEIN (Corylin) FROM YOLK OF IMMUNIZED HENS

**K. Huyghebaert, P. Heyneman, A. Van den Broeck** and **A. Huyghebaert**
Laboratory of Food Technology, Chemistry and Microbiology, Faculty of Agricultural Sciences, University of Ghent, Coupure links, 653 B-9000 Ghent, Belgium

The application of specific antibodies (Ab) in the analysis of foodstuffs is revolutionizing biological research. Such specific Ab are usually produced by blood collection and serum separation from immunized mammals. A simple alternative to this conventional antiserum production is the generation of Ab in chicken eggs. Because of the transfer of Ab from the blood circulation of hens to the egg yolk, an easy collection of the eggs is enough to obtain a continuous source of Ab. These Ab-levels are the same as or often higher than those in the sera of hens. The Ab in the yolk are exclusively of the IgG class and are often called IgY.

Although there exist a lot of methods for the detection of raw hazelnut protein, each of these methods fails in the accurate detection of roasted hazelnut protein in hazelnut spread. Therefore it was the purpose of this research to solve this problem by the generation of chicken IgY raised against denatured hazelnut protein.

Hazelnut protein was extracted from raw and roasted nuts by the procedure of Klein *et al.* (1985). Hens were immunized 3 times at intervals of 2 weeks by injecting natural-state or denatured protein. One month after the first immunization, eggs already contained Ab to hazelnut protein. Ab were extracted from yolk by the procedure of Kint *et al.* (1987) and tested by the gel double diffusion (GDD) technique of Ouchterlony (see Catty & Raykundalia 1988) and the radial immuno-diffusion (RID) of Mancini (see Catty & Raykundalia 1988).

Both the polyclonal sera against natural-state (NS) and denatured (DS) hazelnut protein were tested on their activity by the gel-diffusion methods and the results were very similar. However, the clearest precipitation reactions were obtained by using the NS. 12 hazelnut spreads were subjected to GDD. It became clear that 11 contained hazelnut protein. By means of RID, which is more sensitive than GDD, it could be demonstrated that all the spreads contained hazelnut protein in different amounts. Both the NS and the DS showed no cross-reactivity with milk or almond proteins. Only when the NS was used, was cross-reactivity obtained with walnut protein. This problem can be solved by using DS.

### References

Catty, D. & Raykundalia, C. (1988) In: D. Catty (ed.) *Antibodies*, Vol. 1 — *A Practical Approach*, IRL Press.
Kint, J. A., Huys, A. & Leroy, J. G. (1987) *Archives Internationales de Physiologie et de Biochimie* **95** (5) B215.
Klein, E., Baudner, S. & Günther, H. O. (1985) *Zeitschrift für Lebensmittel Untersuchung und -Forschung* **180** (1) 30–35.

## VI.9  IMMUNOASSAYS FOR THE DETERMINATION OF GROWTH STIMULATOR RESIDUES IN FOOD

**G. Maghuin-Rogister, P. Gaspar, G. Degand, P. Schmitz** and **M. Vandenbroeck**
Laboratoire d'Analyse des Denrées Aliamentaires d'Origine Animale, Faculté de Médecine Vétérinaire, Université de Liège, Bât.B-42 Sart-Tilman B-4000 Liége, Belgium

Treatment of meat-producing animals by veterinary drugs such as anabolizing and repartitioning agents can lead to the presence of residues potentially toxic for the consumer. Identification of treated animals by analyzing their tissues or excreta is not an easy task: on the one hand, the budgets of meat inspection departments are rather limited, while on the other hand, consumers are asking for efficient and numerous controls. A strategy with the best efficiency/cost ratio must be adopted. Critics of the control of residues in the environment and in foodstuffs level the charge that too few chemicals are monitored, the number of controlled samples is not sufficient to ensure detection of contamination, and delays between sample collection and communication of results are too long. These drawbacks are mainly due to the cost, sophistication and time involved in multiresidue analytical methods. Treatment of numerous samples in trace analysis under screening conditions is often difficult with traditional chromatographic methods (TLC, GC, HPLC) even when some of these methods are coupled with spectrometric detection such as mass spectrometry.

Immunoassays offer numerous advantages for the monitoring of environmental contaminants and residues in foodstuffs: low cost, simplicity, low limit of detection, handling of large series of samples, rapidity. There is a growing number of publications and meetings dealing with this topic, and proliferation of companies marketing immunoassays for environmental and food residue analysis.

During the last ten years, we have developed radio-immunoassays (RIA) and more recently enzyme-immunoassays (EIA) for the detection and quantitative analysis of residues of anabolic hormones and repartitioning agents ($\beta$-adrenergic agonists) in tissues and excreta (urine, faeces) from illegally treated animals. The EIA method is based on the competition for a limited amount of a specific antibody, immobilized on plastic, by a constant amount of enzyme-labelled hapten and hapten (drug residue) present as analyte in the sample under examination. The enzyme used is horseradish peroxidase (HRP), and enzymatic activity bound to the plastic after a first incubation is measured by colorimetry after a second incubation in the presence of enzyme substrate (orthophenylene diamine, hydrogen peroxide).

Using EIA, the concentration of artificial anabolics (DES, methyltestosterone, nortestosterone, trenbolone in the sub-part per billion range ($<\mu g/L$) in diluted urine or in extract from solid samples (tissue, faeces) can be determined. In urine, it needs a sample size corresponding to $2.5\,\mu L$ of undiluted urine and it can be performed in less than 2 h on 10–80 samples.

In the future, progress in this technology may allow immunoassay to be conducted *in situ* (farm, slaughterhouse). Nevertheless, positive samples detected by EIA have to be confirmed by a direct spectroscopic method such as infrared or mass spectrometry in order to avoid false positive results.

## VI.10   APPLICATION OF GC/MS TO THE DETERMINATION OF NORTESTOSTERONE AND METHYLTESTOSTERONE RESIDUES IN EDIBLE TISSUES

**C. Van Peteghem, E. Daeseleire, L. Van Look** and **A. De Guesquière**
Laboratory of Food Analysis, University of Ghent, Harelbekestraat 72, 9000 Ghent, Belgium

The illegal use of anabolic steroids in livestock breeding has assumed enormous proportions the last few decades. Measures are taken by the Health Authorities to protect the consumer from possible harmful effects resulting from the consumption of contaminated meat or meat products. Detection of the illegal use of anabolics is difficult because the levels at which they occur in meat are very low (sub-ppb range).

Thirty-four meat samples obtained from the retail trade are analysed for the presence of residues of nortestosterone and methyltestosterone.

The sample extraction and clean-up can be summarized as follows: enzymatic digestion with a proteolytic enzyme, liquid–liquid extraction, pre-purification on disposable $C_{18}$ solid-phase extraction columns and purification by means of reversed-phase HPLC. The appropriate fractions or windows are collected and, after evaporation, subjected to derivatization prior to gas chromatography. For nortestosterone a heptafluorobutyrate ester is selected, whereas methyltestosterone is converted into the corresponding di-TMS derivative. The gas chromatographic as well as the mass spectrometric properties of these derivatives were shown to be the most suitable.

A Hewlett Packard Model 5890 gas chromatograph, equipped with a 25 m HP Ultra-2 (5% phenyl, methylsilicone) fused silica capillary column (0.2 mm I.D., 0.33 μm film thickness) and coupled to a HP Model 5970 mass selective detector is employed for the final detection. The carrier gas is helium at a flow rate of 0.47 ml/min. The injection temperature and the transfer line temperature are 280°C. The oven temperature is programmed from 200 to 280°C at 5°C/min and is kept constant at 280°C during 10 minutes. The injector used is a falling needle system.

Both gas chromatographic retention data and mass spectral data are used for detection and identification. The retention times, expressed in methylene units, are 24.38 for nortestosterone and 28.26 for methyltestosterone. The ions selected for nortestosterone are 666, 453 and 306; for methyltestosterone 446 and 301. Eight of the meat samples are found positive for nortestosterone and only one for methyltestosterone.

It can be concluded from these results that now, more than ever, it is necessary to check for the presence of anabolic steroid residues in meat, which is consumed, rather than in urine samples or injection sites, which are only indirect proofs of illegal administration.

## VI.11   ANALYYSIS OF BETA-AGONISTS IN ANIMAL TISSUES BY GAS CHROMATOGRAPHY–TANDEM MASS SPECTROMETRY

**L. Leyssens, C. Driessen, A. Jacobs, J. Czech and J. Raus**
Dr. L. Willems-Instituut, Universitaire campus, B-3610 Diepenbeek, Belgium

Beta2-receptor agonists (βRAs) are frequently used in man and animals for the treatment of chronic, obstructive pulmonary disease and bronchospasm. Increased protein synthesis and lipolysis are well-known side-effects, which make βRAs also very useful as growth-promoting agents in animal breeding. Although clenbuterol was probably the first βRA to be used for this purpose, at least some of the analogues were found to be potent substitutes. Therefore, a reliable screening procedure which detects βRAs at the low ppb level is needed for an efficient meat control. A highly specific and sensitive method for the simultaneous detection of 7 beta2-receptor agonists in bovine liver homogenates and urine was developed. 10 g of liver was homogenized and treated with Subtilisin A$^{®}$. The resulting enzymatic digest was extracted with t-butanol-ethylacetate (3:7) whereafter the crude extract was purified on a 6 ml Bakerbond$^{®}$ alumina neutral disposable extraction column. Subsequently, the hydrous eluate from the alumina column was buffered at pH 6 and loaded on top of a preconditioned 3 ml Bond-Elut Certify$^{®}$ column. Urine was buffered and loaded onto a 3 ml Certify$^{®}$ column without pretreatment. The analytes were eluted with dichloromethane-isopropanol (8:2) containing 2 per cent ammonia. Ritodrine was added as an internal standard. The obtained extract was trimethylsilylated and analysed on a Finnigan TSQ 70 GC–MS–MS instrument using multiple selected reactions monitoring. The instrument was operated in the positive chemical ionization mode with isobutane as the reagent gas. Multiple selected reactions monitoring was used to confirm the presence of the βRAs. At different time intervals, depending on the retention times of the analytes, the appropriate $[M+H]^+$ ions were selected as parent set masses in the first quadrupole. Collision-induced dissociation was achieved by using argon as the collision gas. Positive identification of a βRA was based on: (1) the retention time, (2) the presence of a typical $[M+H]^+$ ion (parent ion) and (3) the ratio of two or three typical fragments (daughter-ions) generated by collision-induced dissociation of the corresponding $[M+H]^+$ ion. In liver homogenates, the limits of detection were 0.5 ppb for clenbuterol, salbutamol, tulobuterol and mabuterol, 2 ppb for terbutaline, 5 ppb for fenoterol and 8 ppb for orciprenaline. In urine, limits of detection were estimated at least two or three times better.

In conclusion, GC–MS–MS was proven by this work to be an excellent technique for multiresidue analysis of βRAs. It was found to be superior to GC–MS owing to the near-absolute selectivity obtained by the tandem-MS system.

## VI.12 OPTIMIZATION AND AUTOMATION OF THE SPE-CLEANUP AND ON-LINE HPLC DETERMINATION OF N-METHYLCARBAMATE PESTICIDES IN FRUIT AND VEGETABLES
**A. De Kok, M. Hiemstra** and **C. P. Vreeker**
Governmental Food Inspection Service, Pesticides Analysis Department, Alkmaar, The Netherlands

Recently [1] our group has developed a major improved cleanup method for the well-known HPLC determination of N-methylcarbamate pesticides in crop samples [2]. The cleanup method was based on solid-phase extraction (SPE) on aminopropyl-bonded silica columns, using the normal-phase mode. In a subsequent study [3], we applied solid-phase derivatization for the post-column hydrolysis of the carbamates into methylamine, thereby replacing the conventional base hydrolysis with NaOH, prior to the OPA derivatization and fluorescence detection. Besides a strong anion exchange material, which has already been described in the literature, for the first time, magnesium oxide was introduced as a solid-phase catalyst, and the scope of application of heterogeneous catalysis has been extended to the complete group of carbamate pesticides and their metabolites. Major advantages of magnesium oxide over the anion exchange material appeared to be the lower cost of catalyst material, the easier way of packing the post-column reactor and the possibility of executing gradient elutions during a prolonged time of routine analysis.

Finally, the last step in the optimization of the carbamate analysis method has been completed. In the present study the total automation of the SPE cleanup and on-line HPLC determination will be shown. The automation was performed with the commercially available ASPEC (Automated Sample Preparation with Extraction Columns), although some adaptations of the apparatus were required to allow the inclusion of an evaporation step, which is necessary for the change from a normal-phase mode SPE to a reversed-phase HPLC separation.

The application of the totally optimized system in routine analysis of fruits and vegetables will be shown and recovery and reproducibility data will be given. Also, residue data from routine analysis during the last two years are reported.

### References
[1] de Kok, A., Hiemstra, M. & Vreeker, C. P. (1987) *Chromatographia* **24** 469.
[2] Krause, R. T. (1985) *J. Assoc. Off. Anal. Chem.* **68** 726.
[3] de Kok, A., Hiemstra, M. & Vreeker, C. P. (1990) *J. Chromatogr.* **507** 459.

## VI.13　THE USE OF POLYMERIC STATIONARY PHASES FOR THE HPLC DETERMINATION OF BENZIMIDAZOLE-TYPE FUNGICIDES

**A. De Kok, M. Hiemstra, J. A. Joosten** and **C. P. Vreeker**
Governmental Food Inspection Service, Pesticides Analysis Department, Alkmaar, The Netherlands

Among the most employed types of fungicides on fruits and vegetables, the group of benzimidazoles finds ever-increasing application. Owing to the polarity and thermal instability of these fungicides, among which are benomyl, carbendazim, thiabendazole and thiofanate-methyl, analysis of gas chromatography is not preferred. In the last decade, high-performance liquid chromatography has been shown to offer definite advantages for the analysis of the benzimidazoles. However, the chromatographic behaviour of the fungicides on silica-based stationary phases is not ideal and the addition of an ion-pair reagent or a strong (amine) base is required to suppress peak tailing due to adsorption of, in particular, thiabendazole to the free silanol groups of the silica base. The use of mobile phases with a high pH ($>8$), on the other hand, is anything but beneficial for the duration of life of the stationary phase.

For the HPLC analyses with the use of extremely mobile phase conditions, the appearance of stationary phases on polymeric basis offers new possibilities. Besides the well-known polystyrene-divinylbenzene phases, more new polymeric phases are now being developed to circumvent the up till now inherent disadvantages of these phases as to peak asymmetry and the low number of theoretical plates.

In the present study, the applicability of various polymeric stationary phases for the HPLC separation of the benzimidazoles is evaluated and critically compared with the performance of conventional stationary phases. Preliminary results of the inclusion of the optimal HPLC separation in the total fungicide analytical method for fruits and vegetables will be discussed.

## VI.14   A SIMPLE AND RAPID HIGH-PERFORMANCE LIQUID CHROMATOGRAPHIC (HPLC) METHOD FOR THE DETERMINATION OF NITRATES IN VEGETABLES

**H. Beernaert** and **W. van Haver**

Institute of Hygiene and Epidemiology, J. Wytsmanstraat 14, B-1050 Brussels, Belgium

The following statutory regulation has been in force for nitrate in green vegetables since 1989: the nitrate content in lettuce should not exceed $3000\,mg\ NO_3^-\ kg^{-1}$ between 1 April and 31 October and not exceed $4000\,mg\ NO_3^-\ kg^{-1}$ between 1 November and 31 March. The nitrate content in spinach should not be more than $2500\,mg\ NO_3^-\ kg^{-1}$ between 1 April and 31 October and not more than $3500\,mg$ $NO_3^-\ kg^{-1}$ between 1 November and 31 March. Furthermore, the nitrate content in corn-salad, endive, celery and leafy celery should not exceed 3500, 2000, 4500 and $5000\,mg\ NO_3^-\ kg^{-1}$ respectively.

To check these established levels our laboratory has developed a quick, accurate and precise HPLC method. Samples are extracted in a medium of methanol–water and after dilution immediately analysed by ion-pair high-performance liquid chromatography and UV detection at $210\,nm$. To obtain a good peak shape of the eluted nitrate peak, tetrabutylammonium is used as counter-ion. Furthermore, the performance of the analytical column is very important to separate the nitrate peak from the other negative ions, such as $NO_2^-$ and $Br^-$.

Samples of vegetables fortified with 1000, 2000 and $5000\,mg\,kg^{-1}$ shows recoveries of 100.75, 99.35 and 99.03 per cent with variation coefficients of 4.16, 3.42 and 2.67 per cent. At the level of $4000\,mg\,kg^{-1}$, tests of repeatability on lettuce result in an average value of $4125\,mg\,kg^{-1}$ with a variation coefficient of 2.81 per cent. A correction coefficient of 0.987 is obtained comparing the extraction methods using methanol–water and boiling water. Finally, we observe that the nitrate content in the methanol–water extract is stable for several days.

## VI.15   HPLC DETERMINATION OF BIOGENIC AMINES AND EVALUATION OF RESULTS

M. C. Spanjer, T. J. F. Bruin and B. A. S. W. van Roode
Governmental Food and Commodities Inspection Service, Burgpoelwaard 6, 1824 DW Alkmaar, The Netherlands

A liquid chromatographic procedure is developed for the determination of histamine, putrescine and cadaverine in fish (products) and some other foods. The sample cleanup consists of an extraction step with water, followed by precipitation of proteins with trichloro-acetic acid. Samples are chromatographed in a reversed-phase system with a dimethylsulphoxide/water eluent, which also contains dodecyl-sulphate, acetate buffer and ninhydrin. Di-aminohexane is used as an internal standard. Reaction takes place in a stainless steel coil at 150°C, after which the products are detected at 546 nm. Since ninhydrin is dissolved in the mobile phase, no extra reagent pump is needed for post-column derivatization. The repeatability in fish samples containing 100 mg kg$^{-1}$ of each amine was 3.4 mg kg$^{-1}$ for histamine, 4.3 mg kg$^{-1}$ for putrescine and 2.0 mg kg$^{-1}$ for cadaverine (10 determinations). Calibration curves were linear in the 25–400 mg kg$^{-1}$ range (r=0.9996). Recoveries ranged between 80 and 109 per cent. The limits of determination were 29 mg kg$^{-1}$ for histamine, 24 mg kg$^{-1}$ for putrescine and 25 mg kg$^{-1}$ for cadaverine. The usefulness of the method was demonstrated for fish (products), cheese, sauerkraut, wine and meat samples. A comparison of these measurements was made with literature values. In this evaluation, results of 869 fish (products), 553 cheese and 114 sauerkraut samples were involved. It is suggested that, for regulation purposes, in fish (products) the sum of histamine, putrescine and cadaverine should be limited to 300 mg kg$^{-1}$ . In cheese samples tyramine has also to be accounted for, resulting in the suggestion that the sum of tyramine, histamine, putrescine and cadaverine should not exceed the amount of 900 mg kg$^{-1}$. The Mietz-index is considered to be less useful for regulation purposes because it is a relative measure. With respect to health considerations we are more interested in the acceptable daily intake of the total amount of biogenic amines, because of their synergistical action. It will also be demonstrated that from an analytical point of view the index is less accurate as an indicator for the decomposition of food.

## REFERENCES

Joosten, H. M. L. J. (1988) Thesis, Agricultural Univ. Wageningen.
Janssen, F. W. *et al.* (1988) *De Ware(n)-Chemicus* **18** 54.
Spanjer, M. C. & Bruin, T. J. F. (1989) *De Ware(n)-Chemicus* **19** 198.

## VI.16  THE MIGRATION OF DEHP FROM PVC PIPELINES INTO WHOLE MILK

**R. Van Renterghem**
Government Dairy Research Station, Melle, Ministry of Agriculture, Belgium

DEHP (di-2-ethyl-hexylphthalate), also called DOP (di-octyl-phthalate) is widely used as a plasticizer in PVC (polyvinylchloride). PVC pipelines are frequently used in dairy farms and dairy plants. The migration of DEHP from two types of PVC pipelines into whole milk was investigated. Owing to the lypophylic properties of DEHP, migration into skim milk cannot be detected. Forty airtight glass containers of 210 ml, each containing 30 ml of whole milk, were sterilized by autoclave (15 min; 120°C). Rings were cut from the two PVC pipelines. The rings were dipped in a hydrogen peroxide solution and to each of the 20 containers one PVC ring of type A or type B was added under sterile laminar-flow conditions. Five containers with a ring of type A and five with a ring of type B were stored at 4°C. The same was done at 7°C, 20°C and 37°C. After 1, 2, 3, 6 and 8 days of contact time, the DEHP content of the milk in one container of each of the indicated test temperatures and for both types of PVC was determined according to the method of Ruuska *et al.* (*J. Fd. Prot.* (1987) **50** 316–320).

After 1, 2, 3, 6 and 8 days of contact, the DEHP content of the milk in mg $L^{-1}$ and per $dm^2$ contact surface were determined for PVC type A:

    32, 35, 41, 48  and 54 at 4°C
    32, 34, 35, 45  and 57 at 7°C
    42, 69, 79, 110 and unknown at 20°C
    69, 96, 174, unknown and 249 at 37°C

For PVC type B the corresponding concentrations were:

    11, 15, 18, 20  and 30 at 4°C
    17, 19, 23, 33  and 37 at 7°C
    26, 50, 69, 104 and 123 at 20°C
    97, 101, 126, 156 and 197 at 37°C

As contact times are unusually long, the indicated DEHP contents of the milk will never be reached under normal circumstances. On the other hand, swirling the milk in the containers during the migration tests, and hence renewing the liquid in contact with the PVC surface, would increase the migration levels. Since some uncertainties exist about the toxicological effects of DEHP on man, the DEHP content of PVC materials in contact with fatty foods should be restricted.

## VI.17   IDENTIFICATION OF ORGANIC COMPOUNDS IN RUBBER NIPPLES

**J. B. H. van Lierop**
Inspectie Gezondheidsbescherming, Keuringsdienst van Waren Utrecht, The Netherlands

In Utrecht, packaging materials and food utensils, notably those manufactured from organic base compounds, are examined for compliance with the Packaging and Food Utensils Regulation (*Verpakkingen en Gebruiksartikelen Besluit*, VGB). This regulation forms a part of the Dutch Food and Community Act (*Warenwet*).

In this poster a method for testing the migration of compounds from the packaging materials to the food is given. This method includes the identification of organic compounds migrating from the polymeric material.

Full testing of a food contact material for compliance with the VGB, which means examination of the material, composition, volatiles content, migration of constituents and sensory quality is a labour-intensive method. In practice, therefore, the following method is carried out [1]. The material is extracted with diethylether or acetone for rubber nipples, in a ultrasonic bath. The acetone contains three hydrocarbons, C12, C20 and C24, as internal standards. The extract is injected in a gas chromatograph and a gas chromatograph–mass spectrometer combination with capillary columns. By determination of retention indices and comparing with those of reference compounds and by library search using a dedicated spectra library, identification can be carried out.

In 1989, forty-nine rubber nipples available on the Dutch market were investigated. Many compounds not mentioned in the positive list of compounds for the manufacturing of rubber nipples were found. Examples of these not-allowed compounds are:

— dibenzylamine
— acetophenone
— zinc dibenzyldithiocarbamate
— 4,4'-thio-bis(2-tert,butyl-5-methylphenol)
— bis(2-hydroxy-3-tert,butyl-5-ethylphenyl)methane.

Only those compounds are mentioned which were present in a quantity more than half of the quantity of the internal standards.

The analysis of one specific nipple is described in more detail. The chromatogram of this nipple shows the presence of the accelerator zinc diethyldithiocarbamate. The results of the library investigation with the dedicted library are given. In this nipple 10 000 microgram/kg diethylamine, a nitrosable compound was found, an amount 500 times the Dutch level for the total of nitrosable compounds given in the VGB.

### Reference
[1]  Van Battum, D. & van Lierop, J. B. H. (1988) Testing of food contact materials in the Netherlands. *Food additives and contaminants* 381–395.

## VI.18  OCTOBER/NOVEMBER 1989: LEAD TOXICATION OF COWS
**P. A. de Lezenne Coulander**
Inspectorate for Health Protection, Oostergoweg 2, 8932 PG Leeuwarden, The Netherlands

Cattle feed comes from all over the world. In the transportation and blending of the raw stocks to final products things may go wrong. In this case a raw cattle feed material was mixed with zinc and lead ore. The resulting mixture should have been treated as chemical waste, but was used as a blending component for cattle feed. This feed was delivered at the beginning of October 1989 to about 300 farmers in the provinces of Groningen and Friesland (The Netherlands) and to the United Kingdom. Some 15 000 cows were affected in The Netherlands alone, of which about 100 died. The Inspectorate for Health Protection was informed about the case history of the cows on October 30, 1989.

### The consumer needed protection
In The Netherlands the following authorities were involved.

The regional Health Service for Animals was consulted by veterinarians in the affected area. Many blood, feed and organ samples were analysed to enable a diagnosis for the disease.

From the Department of Agriculture, the General Inspection Service (AID) conducted an investigation concerning the origin of the feed and controlled, together with the Inspection Service for Meat, Animals and Animal Products (RVV), the slaughter of affected animals.

The Veterinary Public Health Inspectorate (VI), part of the Ministry of Welfare, Health and Cultural Affairs, inspected the RVV activities with respect to the animals and meat.

The Inspectorate for Health Protection, also a part of the latter Ministry, started to coordinate the above-mentioned authorities and took measures to avoid the polluted raw milk being used for consumer products such as milk and cheese.

The produced polluted milk was separated from the non-polluted milk and converted to butter and milk powder. These two products were kept in stock, awaiting further decisions.

The analytical data were produced using flameless atomic absorption spectrometry (Perkin–Elmer 5100 with Zeeman background correction) after destruction in PTFE bombs. Highest analytical data obtained in various products tested, in microgram lead per kilogram sample, are:

| | | | |
|---|---|---|---|
| Raw milk | 330 | Kidneys | 15000 |
| Milkpowder | 1370 | Liver | 5300 |
| Cheese from farm | 2000 | Muscle tissue | 190 |
| Buttermilk | 220 | Blood | 110 |

The above products were kept away from the human food market.

The biological half-life of lead in milk appeared to be just less than four days. Therefore the effects of this disaster diminished within two weeks. (The Dutch Food Law allows a lead content of maximal 50 micrograms per kilogram milk.)

Owing to the effectiveness of the coordinated activities it was possible to keep almost all polluted products away from the market. Only in the last weeks of October 1989 a few cheese-producing farmers, unaware of this case, sold *and* consumed polluted products.

## VI.19   ALUMINIUM INTAKE — ALZHEIMER'S DISEASE?
**H. H. S. Roomans, E. J. M. Konings** and **W. H. P. Franssen**
The Inspectorate for Health Protection in The Netherlands, Keuringsdienst van
Waren, Maastricht, Florijnruwe 111, 6218 CA Maastricht, The Netherlands

In recent editions of *The Lancet* questions were raised such as: 'Does the aluminium
contamination of milk formulas pose a risk to the health of infants?' and 'Should soy
formulas not be recommended without specialist advice?'

There is therefore growing public concern that aluminium may be the cause of
dementia, especially Alzheimer's disease. With reference to these articles we have
analysed 134 products (infant formulas and baby-foods) consisting of 268 samples.

The aluminium content of infant formulated and baby-foods is measured by
graphite furnace atomic absorption spectrometry after previous wet acid digestion.
Soy protein is the component of these foods with the major aluminium
concentration.

To get a picture of the total dietary aluminium intake during the first year of life
we made calculations based on:

— the mean aluminium level as analysed in the foodstuffs
— the recommended amount of foodstuff as given on the label
— the composition of the total diet for infants from 0 to 12 months based on
   information from the Dutch Infant Welfare Centre (*consultatiebureau*)
— the mean aluminium content of Dutch tap water (<10 µg/L) is not negotiated in
   the calculation, because the change in aluminium content is marginal.

The amount of aluminium consumed per day by the different age-groups, for
lowest and highest aluminium level is:

0–3   months: 0.20 to 0.90 mg per day
3–6   months: 0.10 to 1.50 mg per day
6–9   months: 0.10 to 1.30 mg per day
9–12 months: 0.10 to 0.30 mg per day

The daily (even the maximum calculated) intake of aluminium for infants fed
with infant formulas and baby-foods for a period of 12 months is below the ADI.

The conclusion is that, in the first year of life, at no time is the allowable daily
intake (ADI) of aluminium in nutrition exceeded.

## VI.20 TWO APPLICATIONS OF THE ISOTACHOPHORETIC SEPARATION TECHNIQUE IN FOOD AND FOOD PRODUCTS

**K. Stikwerda, A. M. Westra** and **R. Fransen**
Inspectorate for Health Protection, PO Box 465, 9700 AL Groningen,
The Netherlands

### Principles of isotachophoresis

Isotachophoresis (ITP) is the electrophoretic separation of ionic species according to mobility, which is achieved in a capillary column under conditions of constant current.

The ITP experiment consists in taking a sample containing a mixture of ions of varying mobilities in a capillary at the interface between an electrolyte containing an ion of high mobility (LE) and an electrolyte containing an ion of low mobility (TE), all three having a common counter-ion. When the appropriate electric field is applied to the system the sample ion of the highest mobility will move faster than ions of lower mobility. The net effect will be that the ions will be separated according to their mobilities and then continue to move along the capillary.

### Isotachophoretic determination of l-histamine in fish and fish products

An isotachophoretic assay has been developed to determine the $l$-histamine content in fish and fish products. Sample pretreatment is simple, consisting of extraction with water in the presence of a few drops at diluted hydrochloric acid, and filtration.

The clear filtrate is submitted to isotachophoretic analysis (*conditions*: leading electrolyte: 5 mM dimethylarsenic acid sodium salt (trihydrate) in water +0.4% hydroxypropylmethylcellulose; terminating electrolyte: 5 mM $l$-histidine, adjusted with hydrochloric acid to pH=4.4; current: initial strength $100 \mu A$ till 20 kV is attained, then reduced to $50 \mu A$; capillary: 63 cm teflon, i.d. 0.5 mm; temperature: 10°C; detector: conductivity detector. A special 'isotachophoretic' integrator was used for data acquisition. The contents of the leading- and terminating-electrolyte vessels were replaced after each injection (this is essential for the procedure).

A linear relationship was found between $l$-histamine standard solutions ranging from 0 to 5.3 mg/100 ml (r=0.9988; $Y=145 X-10.7$). The recoveries were 99.7, 97.5 and 93.5 per cent for added amounts of 100, 120 and 200 mg/kg histamine in fish (mackerel). The coefficient of variation for the complete procedure was 4.46 per cent (level 100 mg/kg histamine; n=7 determinations).

### Isotachophoretic determinations of glutamic acid in meat and meat products

The determination of glutamic acid in meat and meat products by isotachophoresis is described. Treatment of laboratory sample consists of extraction of sample with water, and filtration. The clear filtrate is submitted to isotachophoretic analysis (*conditions:* leading electrolyte: dissolve 200 mg arkopal N-100 in 90 ml water, add 10 ml 0.1 HCl, and adjust the pH to pH=3 with β-alanine; terminating electrolyte: 0.005 M hexanoic acid in water; current: $100 \mu A$; capillary: 25 cm teflon, i.d. 0.55 mm; temperature: ambient; detector: conductivity detector).

A linear relationship was found between detector signals (zone length) and a series of glutamic acid standard solutions ranging from 0 to 45 mg/100 ml (r=0.99975; $Y=0.505 X+0.67$). The average recoveries were 101.4 per cent, 95.7 per cent and 95.5 per cent for added amounts of 0.10, 0.20 and 0.30 per cent glutamic acid in meat-products. The coefficient of variation for the complete procedure was 2.7 per cent (level 0.13 per cent glutamic acid; n=7 determinations). The proposed procedure proved to be fast and suitable for routine analyses yielding reliable qualitative and quantitative results.

## VI.21   GALLOTANNINS IN PROCESSED FOODS: A TECHNOLOGICAL AND ANALYTICAL EVALUATION

**Roger Mussche, Jan Bijl** and **Christian De Pauw**
Omnichem NV, Wetteren, Belgium

Gallotannins have been used for more than a hundred years for stabilization and clarification, for structure improvement and flavour enhancement, both in liquid and solid foods.

The new high-molecular and highly purified gallotannins can be added to liquid using an on-line method with instantaneous reaction. Final residues of the gallotannins in, for example, beer, champagne, wine and fruit juices are below the detection limits of the classical analytical techniques. An adapted flow sheet gives the possibility for a proportional, continuous injection in the liquid stream, followed by a filtration of the protein–gallotannin complex on an adequate filter system. The filter aid should be of the appropriate particle size to retain the precipitate.

Small amounts of gallotannins are added to solid foods, e.g. jelly with gelatin, to improve the structure. In certain dressings they can be used to enhance the flavour.

The different analytical techniques which are used both as technical aids and as control methods are described herewith.

A combination of reversed-phase HPLC and an in-line rapid-scanning detection system enables the analyst to demonstrate the identity and the purity of the gallotannin.

Beverages can be tested for the eventual presence of ppm amounts of residual gallotannin by enzymatic hydrolysis with tannase, and subsequent RP–HPLC of the produced gallic acid.

Different methods can be applied to determine the required amounts of gallotannin for stabilization and clarification. Colour reactions between the excess of proteins and, for example, Coomassie blue or Amido black are used both for optimization studies and for control after addition.

The amounts used in solid foods are analysed by solubilization in acetone/alcohol of the gallotannin fraction, precipitation of the proteins, and filtration, followed by either direct HPLC or determination of the gallic acid content after enzymatic hydrolysis.

## VI.22 SIMULTANEOUS DETERMINATION OF ASCORBIC ACID AND ERYTHORBIC ACID IN MEAT PRODUCTS

**Willy Schüep** and **Elfriede Keck**
Department for Vitamin and Nutrition Research, F. Hoffmann-La Roche Ltd,
CH-4002 Basle, Switzerland

Oxidation is one of the major reactions in connection with undesirable changes in food products. Antioxidants have therefore a significant function in stabilizing foodstuffs. Ascorbic acid has excellent antioxidant properties and it is utilized in different product forms and processes of the food industry. Erythorbic acid is an isomer of ascorbic acid and its use as a food additive is not legally permitted in the European Community and also in most other European countries. It is therefore important to have a good working analytical procedure which allows the differentiation of these two compounds in food products.

Numerous assay methods have been described for the determination of ascorbic acid. Most of the older procedures do not differentiate between ascorbic acid and the isomeric erythorbic acid. The upcoming technique of high-performance liquid chromatography made it possible to achieve a good and fast separation of these two compounds. An easy and simple sample treatment allows the processing of a high number of samples. The HPLC system used consisted of a reversed-phase column, a mobile phase containing an ion-pairing agent and UV-detection.

A series of meat products, such as ham, sausages, frankfurters, pâté etc., from 14 different countries has been assayed. Out of 83 samples bought on the market, 20 were found with no detectable amounts of ascorbic acid. 63 samples had ascorbic acid levels up to 472 ppm and 9 were identified as containing erythorbic acid. These data reveal that more than 10 per cent of the meat products analyzed were processed using a compound not found in natural food.

## VI.23   ISOELECTRIC FOCUSING OF GENETIC VARIANTS IN COW'S MILK

**H. De Moor** and **A. Huyghebaert**
Committee for Scientific and Technical Research in Dairying, Laboratory of Food Technology, Chemistry and Microbiology, Faculty of Agricultural Sciences, University of Ghent, Coupure links, 653 B-9000 Ghent, Belgium

The manufacturing properties of milk are influenced by many factors, e.g. protein content and composition. The genetic variants (polymorphism) of the proteins in cow's milk also appear to have an influence on the renneting and heat stability properties of the milk and on cheese yield.

For the phenotyping of milk proteins two methods using isoelectric focusing (IEF) are used. Method A is a modification of the method of B. Seibert, G. Erhardt & B. Senft (*Animal Blood Groups and Biochemical Genetics* (1985) **16** 183) and method B is described by H. Bovenhuis & A. C. Verstege (*Netherlands Milk and Dairy Journal* (1989) **43** 447).

In method A, the milk proteins are separated by precipitation at pH 4.6 in a casein and whey protein fraction. $\alpha_{s1}$-CN, $\alpha_{s2}$-CN, $\beta$-CN, $\kappa$-CN, $\alpha$-La and $\beta$-Lg are separated in thin-layer polyacrylamide gels (1 mm). The protein fractions contain urea and $2\beta$-mercaptoethanol. The pH gradient is formed by ampholytes: pH 2.5–5 and 4–6.5.

In method B, IEF is performed by the Phast System (Pharmacia). Phast Gels IEF are first soaked in a urea solution containing Triton X-100. After incubation, the gels are soaked again in a urea solution containing Triton-X 100 and carrier ampholytes, pH range 4.2–4.9, 4.5–5.4 and 3.5–5. The separation of the milk proteins (whole milk mixed with urea and $2\beta$-mercaptoethanol) is carried out in one single run. After focusing, the gel is fixed, washed, stained with Coomassie R-350 and destained.

Both methods gave good results. The main advantage of method B is the time saving, and the method is now used to investigate the distribution of the genetic variants of milk proteins in milk in the main Belgian cow breeds. This study will be followed by research on the influence of the genetic variants on milk production traits, on the detailed milk composition, on the association between milk components and the technological properties of milk. Knowledge of these influences is necessary if the genetic variants are to be used as a selection criterion in the breeding of cows.

This research has been supported by the Institute for the Encouragement of Scientific Research in Industry and Agriculture.

## VI.24   DETECTION OF COW'S MILK PROTEINS IN DAIRY PRODUCTS FROM GOATS

**P. Van Dael** and **H. Deelstra**
Laboratory of Food Sciences, Department of Pharmaceutical Sciences, University of Antwerp (UIA), B-2610 Antwerp, Belgium

The detection of cow's milk in the milk products of goats was based essentially on differences in fat and protein composition.

For the analysis of the fatty acid composition of the dairy products, gas chromatography is often used. These methods have a high detection limit but are ineffective when skim milk has been used as an adulterant.

Immunological and electrophoretical assays are suitable for the analysis of protein fractions. The immunological technique is faster than conventional electrophoresis, but has the draw-backs that the seroproteins are affected by heat treatment and that specific antisera are rather difficult to obtain. Electrophoresis of the casein fraction has the advantage over the immunological assay in that these proteins are more heat stable, but the electrophoretic patterns obtained are complex.

Advances in high-performance liquid chromatography have made it possible to obtain fast and quantitative separations of milk proteins of different origin on various analytical columns.

In this paper the separations of bovine and caprine caseins on a Mono Q column are described. This method has been applied to the determination of bovine caseins in dairy products from goats.

## VI.25 DETERMINATION OF NON-MEAT PROTEINS IN HEATED MEAT PRODUCTS BY MEANS OF ELISA

**L. Deweghe, L. De Cock, J. Lenges** and **P. Reinquin**
Analytical and Experimental Station, COOVI-CERIA, E. Gryzonlaan, 1, B-1070 Brussels, Belgium

'Non-meat proteins' such as soya protein, casein, whey protein, wheat gluten and ovalbumin are used in meat products because of the way their technological properties improve a lot of physico-chemical paramaters in the final product. Nevertheless, their addition to meat products is regulated by the Belgian Legislation and, consequently, testing laboratories have to devise an adequate method to determine the conformity of these meat products. Moreover, in heated meat products the proteins possibly have undergone denaturation. In view to this problem, a specific method providing a renaturation and a solubilization of the proteins has to be considered. An Enzyme-Linked Immunosorbent Assay or ELISA satisfies these conditions.

A methodology by Ring (1987) for the quantitative determination of soya protein in cooked sausages is worked out and adapted and completed to determine casein, whey protein and wheat gluten as well. The method is an indirect ELISA and includes the following steps:

— extraction and renaturation of the non-meat proteins from samples and standards;
— direct coating of the microtitration plate with these antigens;
— addition of a rabbit antiserum against the protein to be detected;
— addition of the conjugate, anti-rabbit IgG conjugated with peroxidase;
— addition of the substrate 5-aminosalicylic acid with hydrogen peroxide, which, in the presence of the protein to be detected, is converted into a coloured compound; the intensity is measured by spectrophotometry.

The procedure has been tested on cooked sausages with known compositions, heated in a steam case at 80°C until an internal temperature of 70°C is attained. Soya-isolate, casein, whey protein, wheat gluten are added in concentrations of 0.4–1 per cent of the product as such. After optimization (adaptation of the incubation period for coating of the antigens on the plate and selection of the most convenient microtitration plate) a proper and reproducible calibration curve is obtained for each of the four proteins. In general, the recoveries amount to 96–108 per cent for soya protein, 95–110 per cent for casein, and 96–112 per cent for wheat gluten. For whey protein, recoveries of only 70 per cent are found; this could be due to the low protein content (13 per cent) of the whey powder used in the meat preparations, so that the final concentration of whey protein in the sausage is close to the detection limit.

Cross reactions between casein and anti-whey serum and between whey protein and anti-casein serum occur, probably due to the impurity of the used milk proteins.

The detection limit for the different proteins is 0.1 per cent, even lower for wheat gluten.

The whole test is realizable on one microtitration plate; the total duration of the complete analysis takes about one day.

## VI.26   SPECIES IDENTIFICATION OF MEAT AND MEAT PRODUCTS BY TLC–GC

**H. F. De Brabander** and **J. Van Hoof**

Laboratory of Chemical Analysis of Food from Animal Origin, Veterinary Faculty of the University of Ghent, Caseinoplein, 24, B-9000 Ghent, Belgium

Species identification (SPID) may be defined as the search for the origin of the material of which foodstuffs are composed. The determination of the relative proportion of several meat species in one meat product is a complex analytical problem. The reasons for species-identification are various. Next to economical considerations (e.g. differences in price, trade regulations) also religious and ethical motives may lead to species analysis. The methods used in SPID may roughly be divided into three groups: anatomical–histological, immunological and physico-chemical. Chemometric methods, such as pattern recognition, are often used to solve complex data sets.

In this paper an update of a physico-chemical method for the identification of the most important animal fats is described. Identification of animal fat is important, not only for bulk fat but also for the identification of heated meat products. A method has been developed for the extension of the data from the fatty acid composition to those in 2-position. By an enzymatic TLC method the 2-monoacylglycerols (monoglycerides = M) are isolated. The M fraction contains the fatty acids incorporated in 2-position (beta-position) of the triacylglycerols.

The isolation of the 2-monoacylglycerol fraction is performed with an enzymatic 'all on one plate' method. A homogeneous lipase reaction band (ca 1.5 cm×7 cm) is formed on silicagel plates (10 cm×20 cm). Afterwards, 100 μL of a fat solution in hexane is evenly applied over the lipase reaction band. After 10 min incubation at 40°C the plate is dried carefully and the lipid mixture is concentrated into a narrow band by developing the plate three times with diethylether over a distance of ca 5 cm. The lipase reaction band is removed by cutting off that part of the plate. The remainder of the plate is developed with hexane-diethylether (80:20, V/V). After drying, the M fraction is eluted with diethylether. The M and T fraction are derivatized with sodium methylate to FAMEs and analysed by capillary gas chromatography.

By combination of the fattty acid composition in 2-position (= M) and the total fatty acid composition (=T) species specific parameters could be calculated. These parameters may be combined to two-dimensional graphs in which the different species appear as separate clusters. The selection of the 'best' parameters was carried out with a PC (Apple Macintosh) and by pattern recognition.

Species identification of an unknown sample with the aid of a PC is carried out as follows. After TLC–GC analysis all identification parameters of the sample are calculated with a computer program, designed at our laboratory. The unknown data set is then plotted on a first cluster graph. This graph gives the best separation between pork fat and the other fats. Depending on the position of the unknown data pair on this graph, different strategies are followed:

— If the point, formed by the unknown data pair falls within the cluster of pork fat or another fat (beef, horse or chicken fat) the fat sample is to be considered as pure. Other graphs could be used for obtaining more information. With another computer program it can be calculated if the unknown falls within the 95 or 99 per cent confidence ellipse of the species cluster.

— If the data point is situated between two clusters, a mixture of (at least) two species is present. The relative amount of both fats in a binary mixture may be estimated from the relative distance from the data point to the centre of the two clusters. With another computer program the mean composition of the 'main' fat is subtracted from the data set for three mixing percentages in the neighbourhood of the estimation. The parameters of these 'secondary' fats are plotted on the cluster diagram giving a straight line which cuts the cluster of the secondary fat. From this cluster the identity of the secondary fat is derived.

All the programs used are written in Microsoft Basic (Apple Macintosh).

## VI.27   ISOLATION AND QUANTITATIVE ESTIMATION OF COOKED FOOD MUTAGENS IN BEEF EXTRACT

**N. Henin** and **C. de Meester**

Université Catholique de Louvain, Département de Pharmacie, Unité de Mutage-nèse et de Tératogenèse, UCL-72.37, Belgium

The heat treatment of proteinaceous foods generates contaminants whose mutagenic and carcinogenic properties are well documented. These compounds belong to the heterocyclic amines family and have been identified in foods like cooked meat and fish or industrial products like beef extract.

In our study, we examined a number of commercial samples of gravies containing beef extract. These were extracted with methylene chloride and the so-called 'basic fraction' (BF) was tested for mutagenicity, following the Ames' test. All the samples containing beef extract were mutagenic and a good correlation was found between mutagenic activity and creatinine content.

The HPLC analysis of the BF of food-grade beef extract on a reverse-phase C18 column with a $KH_2PO_4$ 0.02 M/acetonitrile (90:10) mobile phase (pH=5) revealed the presence of two mutagens: 2-amino-3-methylimidazo [4,5-f] quinoline (IQ) and 2-amino-3,8-dimethylimidazo [4,5-f] quinoxaline ($MeIQ_x$). The quantitative estimation of these two compounds, based on internal standards calibration curves, were: 133 ng/g for IQ and 201 ng/g for $MeIQ_x$. These values are rough estimates however, as the extraction and purification methods need to be improved, particularly for IQ. This proved to be completely unsuitable for the analysis of more complex preparations like gravies.

### *Acknowledgement*

We wish to thank Mr A. d'Adesky (Inspection des Denrées Alimentaires) for the financial support for this study.

### VI.28   COGNAC ADULTERATION

**R. Hittenhausen-Gelderblom, W. B. H. Kennedy, M. van Putten and**
**H. A. van der Schee**

Inspectie Gezondheidsbescherming, Keuringsdienst van Waren, Hoogte Kadijk 401, 1018 BK Amsterdam, The Netherlands

In the past, the Governmental Food Inspection Service detected adulteration by the determination of the quantitative amounts of the fusel alcohols present and the ratio between them. However, nowadays this method does not yield up many offenders. This made us suspect that these parameters had been adjusted by *malafide* manufacturers and prompted us to search for other parameters:

— *GLC–aromagramdata* of di-isopropyl-ether extracts (Likens and Nickerson distillation-apparatus);
— *purge and trap* (headspace injection) for the determination of volatiles.

**Example of an adulteration**

In total, 33 samples of cognacs were analysed. Cognac from only one manufacturer showed the following differences:

- Fusel alcohols      — only a slight difference in the isoamyl–alcohol ratio was detected.
- Aromagram         — the concentration of most volatiles in the gas chromatogram was less than those found in the reference cognacs.
- P&T chromatogram — the amount of ethylheptanoate is too high.
                    From the literature and from our own experience this compound is virtually absent in cognac-type distillates.
                    — a strange unknown peak was detected.

The unknown peak could be found in the reconstructed ion chromatogram detected by GC–MS and identified with a mass-spectrum as di-isoamylether (di-3-methyl-butylether). Moreover:

(1) It appears that the identified ether is an important contaminant in commercial 3-methylbutanol.
(2) In commercial 3-methylbutanol, 2-methylbutanol could also be found.
(3) Ethylheptanoate is a well-known flavour compound, which gives a winey/cognac taste to beverages.

**Conclusion**

The addition of ethylheptanoate and fusel alcohols — to mask the adulteration — can explain all the detected differences.

*The adulteration has been proved.*

## VI.29  APPLICATIONS OF AMINO ACID ANALYSIS IN FOOD CONTROL

**Ooghe W.**
Laboratorium voor Bromatologie, Universiteit Gent, Harelbekestraat 72, B-9000 Ghent, Belgium

Amino acids are present in almost all foods, either in a free form or linked together as peptides, polypeptides or proteins. Often a foodstuff may be characterized by the relative amounts of the amino acids present.

Amino acids may quantitatively be determined by many different (chromatographic) techniques. From our experience, however, cation-exchange chromatography using lithium citrate buffer solutions, followed by the ninhydrin colour reaction is most appropriate for the amino acid determinations, including proline and hydroxyproline, in food products.

Sample preparation depends whether free or total amino acids have to be determined. Mostly free amino acids may be injected directly after dissolution or dilution of the sample in the injection buffer solution of pH 2.2 and after ultrafiltration through a membrane filter. Hydrolysis by boiling under reflux in azeotropic (6M) hydrochloric acid is most suitable for the determination of total amino acids, although tryptophane is completely destroyed.

The applications of the amino acid analysis in food control are nearly unlimited and may roughly be divided into three main groups:

— the determination of the chemical score of food proteins — which is related to their biological and nutritional value — by comparison of the pattern of the essential amino acids to a reference pattern;
— the influence of technological processes and additives on the nutritional value;
— the detection of fraudulent manipulations or economically profitable falsifications, especially in fruit juices, fruit beverages, wines and meat products.

Especially for that last group of applications, the combination with pattern recognition techniques or comparison of the data to appropriate standards by means of a statistical $\chi^2$- or F-test, may provide valuable and reliable information within a short time.

## VI.30   NEW ANALYTICAL POSSIBILITIES IN FRYING-FAT REGULATION

**M. C. Dobarganes, M. C. Pérez-Camino, G. Márquez-Ruiz** and **M. V. Ruiz-Méndez**
Instituto de la Grasa y sus Derivados, Avda. Padre Garcia Tejero, 4, 41012-Sevilla, Spain

The increase in the consumption of fried foods in developed countries has led to regulations limiting the alteration level in frying fats. The most accepted criterion recommends rejecting the fat when its level of polar compounds is higher than 25–27 per cent, the determination being performed by the IUPAC method [1].

As it is known, polar compounds are a complex mixture of alteration products. They are principally the result of the action of atmospheric oxygen, the water content of foodstuff, and high temperature. It is important to note that, while the hydrolysis involves breakage of the ester bond with formation of fatty acids, monoglycerides and diglycerides — the normal compounds originating in the stage previous to the fat absorption — the oxidative and thermal degradations take place in the unsaturated acyl groups of the triglycerides, modifying the nutritional properties of the fat. This means that polar compounds differ not only in polarity or molecular weight but also in physiological significance.

Taking into account that regulation is demanded because of the fact that alteration compounds in abused frying fats may be harmful to the consumer, cut-off levels should be based on the quantitation of the compounds that may impair the nutritional value of the fat.

Two different possibilities are proposed to quantify the compounds with nutritional significance:

— determination of polar compounds coming from thermal and oxidative alteration;
— determination of polar fatty acids.

Separation and quantitation of representative polar compounds or polar fatty acids have been performed by high-performance size exclusion chromatography [2]. The results demonstrate that fats with the same level of polar compounds have different contents in thermoxidized compounds. A very high correlation between the two proposed methods has been obtained.

### References
[1] Waltking, A. E. & Wessels, H. (1981) *J. Assoc. Off. Anal. Chem.* **64** 1329.
[2] Dobarganes, M. C., Pérez-Camino, M. C. & Màrquez-Ruiz, G. (1988) *Fat Sci. Technol.* **90** 308.

## VI.31   INFANA, THE COMPUTERIZED INFORMATION SYSTEM OF THE BELGIAN FOOD INSPECTORATE

**G. Temmerman**

The Food Inspectorate, Ministry of Public Health and the Environment, Brussels, Belgium

The Belgian Food Inspectorate has its Head Office at the Department of Health in Brussels. It runs a network of local branches throughout the nine provinces, manned by food inspection officers working in the field. The Head Office organizes the work of the field inspectors, who carry out the inspection tasks and report back to the Head Office.

In order to improve the information exchange between the headquarters and the local inspection officers, a computerized system has been established within the Belgian Food Inspectorate.

This system consists of a number of databases established by the Head Office (product codes, analyses, inspection tasks, officers, laboratories, etc.) and of databases containing inspection data collected by the officers in the field. Each day, the inspection officers input their inspections, samplings, warnings, etc. to their personal computer. The analytical laboratories transmit the results of the sample analyses to the central computer, either directly in on-line mode or on floppy disks sent to the headquarters by post.

When the inspection officers send their inspection data to the host computer, they automatically receive the outcome of the analyses of the samples they have taken.

The data in the host computer can be consulted by all the members of the service. However, they can be updated only by the person who fed them to the computer. The laboratories have access only to those data that are of interest to them. A statistical program allows the Head Office to process the inspection data statistically.

The VIDEOTEX communication service of the national telephone company (Prestel mode) has been selected as the medium of communication. This system is very user-friendly and can be used by the officers wherever they are; unfortunately, it is rather slow.

In case there are problems with the VIDEOTEX system, the officers can also send their data on floppy disks to the Head Office for a direct file-transfer.

All the officers have a complete personal computer set, including a modem. This means that all the officers are able to carry out their own administrative tasks and to consult remote databases. in addition, there are also several mailing-box systems, which can be used by all the members.

## VI.32   FOOD CONTROL PATTERNS AND ORGANIZATION OF PUBLIC LABORATORIES

**A. De Sa Lopez**
Centro de Investigaciòn y Control de Calidad, INC, Madrid, Spain
**F. Centrich**
Laboratorio Municipal, Ayuntamiento Barcelona, Spain
**I. Eguileor**
Laboratorio de Salud Pùblica, Osakidetza/SVS, Bilbao, Spain

The main organizational aspects of public laboratories for a proper adaptation to the possible food control patterns are described. From the point of view of food analysis the food surveillance or control could be simplified and reduced to two different approaches:

(1) *Final product control.* This system is devoted to the enforcement of regulations on food commodities or quality assessment. It requires a detailed legislation based on an exhaustive knowledge of the product to be analysed. Laboratory resources are in many cases legally imposed, demanding the establishment of a laboratory network with a lot of routine work employing standardized methodology.

(2) *Global surveillance of food safety.* This pattern involves much more resources, sectors, and disciplines. It requires and contributes to an integral knowledge of the foodstuffs from the raw products to the food commodities, not only in the nutritional aspect but also in the safety ones. Special organization and management effort is required for priorities setting, programmes development and activities evaluation. Great versatility with high technological and scientific standards are demanded from the laboratories with a lot of dedication to analytical methodology research and development.

Actually no country has a pure food control pattern as described, but different approaches coexist and are applied to diverse food sectors according to their development, economical significance or potential food safety risk involved. For laboratory organization the characteristics of production, distribution or consumption of foodstuffs require a strict definition of activity fields in the laboratories. This implies a structure based on laboratory specialization at country or even community level because of the internationalization and complexity of the problems related to food control.

The described situation must be considered under the Food Control Directive 89/337 perspective that requires the accreditation and mutual recognition of the food analysis laboratories in order to guarantee integral control systems. To accomplish these goals, the laboratories, in addition to becoming accredited and to using methods validated through collaborative trials, will have an important load of work in continuously demonstrating their aptitude and analytical quality.

This point presents, among others, two main practical aspects: firstly, the circulation of information and the participation in intercomparison or proficiency testing schemes, undertaken by reputable organizations with international agreement of the institutions responsible for food control; and secondly, the development of activities allowing an effective exchange of knowledge and experience among laboratories, assisting continuous scientific updating.

## VI.33   SOME ESSENTIAL MINERALS AND TRACE ELEMENTS OF SUNFLOWER PRESSCAKE MEAL

**Verena Ndoreyaho-Mukankusi**[1,2] and **Alfred Noirfalise**[2]

[1]Université Nationale du Rwanda, Butare, Rwanda

[2]Université de Liège, Faculté de Médecine, Laboratoire de Toxicologie et de Bromatologie, 151, Bd de la Constitution, 4020 Liège, Belgium

Sunflower presscake meal constitutes an interesting supplement to the Rwandan diet, which is essentially of plant origin.

This meal contains the following essential major minerals (in mg/100 g of dry matter for an average of nine sunflower varieties): calcium (219.09), phosphorus (1055.0), sodium (65.52), potassium (1155.66) and magnesium (662.68).

The trace element (iron, copper and zinc) content in these meals can be considered theoretically sufficient to cover the related needs, if their physiological activity aspects are not taken into account at this stage of this research. These trace elements are essential for preventing malnutrition in general and, particularly, for the treatment of protein and energy deficiency, which is endemic in Rwanda. In close relation with this malnutrition, patients also become anaemic. A diet including sunflower presscake meal can easily relieve this disease, considering that its iron content reaches, on average, 28.41 mg/100 g dry matter (average of nine varieties).

Susceptibility to infection in malnutrition can be ascribed to zinc deficiency. The zinc content of these sunflower presscake meals (14.82 mg/100 g dry matter) is theoretically high enough to cover those needs.

Considering the role of copper in human nutrition, its supplementation improves the growth of copper-deficient infants recovering from malnutrition. The copper content of sunflower presscake meal, which reaches 5.52 mg/100 g dry matter (average of nine sunflower varieties), is also high enough to cover those needs.

## VI.34   TOTAL DIET STUDY IN THE BASQUE COUNTRY (SPAIN)

**I. Urieta**
Osakidetza, Direccion de Salud de Bizkaia, Maria Diaz de Haro 60, 48010 Bilbao, Spain
**M. Jalon**
Osakidetza, Direccion de Salud de Alava, Santiago 11, 01004 Vitoria, Spain

The Basque country is a small region in the north of Spain, 7261 km$^2$ in area and with a population of 2.2 million. Between 1988 and 1990 a food survey was conducted by the Health Department of the Basque Government with the purpose of identifying the types and quantities of foods that are consumed by the population. This survey employed '24h-recall' interviews and an individual food frequency questionnaire and is the basis for the determination of the average intake of certain food constituents of concern, through a total diet study.

The definitive total diet study has started recently, but during 1990 a pilot study based on the preliminary results of the food survey was conducted. The purpose of the pilot study was to foresee the possible difficulties which would appear in the main study and also to set up the techniques for the selected substances. It included the sampling, preparation and analysis of three 'market baskets' from three different locations.

The poster shows a detailed scheme of the different stages of the study, which can be summarized as follows. Using the information provided by the food survey, the average diet of the population is established and the major items in the national diet identified. Then the food list is prepared and the food items included (91 categories of foods) are purchased. The foods are prepared as for eating and combined in 16 groups of similar foods for analysis. These groups are: eggs, meat, meat products, fish, milk, dairy products, bread, cereals, pulses and nuts, potatoes, vegetables, fruits, sugar and preserves, fats and oils, non-alcoholic beverages and alcoholic beverages. Each group is then analysed for some of the substances of concern which initially are:

- *Heavy metals:* mercury, lead, cadmium and arsenic.
- *Organochloride pesticides:* hexachlorobenzene, hexachlorocyclohexane ($\alpha$, $\beta$, $\gamma$, $\delta$), DDT and its isomers (DDE, TDE), dieldrin, aldrin, endrin, heptachlor, heptachlor epoxide, endosulphan ($\alpha$, $\beta$), and methoxychlor.
- *Minerals:* zinc, iron and selenium.

Also, in the milk and dairy products groups, aflatoxin $M_1$ is determined.

In the poster we present the results of the analysis of each group of foods, as well as the techniques used and their limits of detection. In summary we can say that, even assuming residues below the limit of detection of the method to be at that level, the estimated total intakes for the chemical contaminants are well below the acceptable daily intakes.

## VI.35   INSECTS IN FOOD AND FOOD PRODUCTS
### V. C. van Woerkom
Inspectorate for Health Protection, Postbus 465, 9700 AL Groningen, The Netherlands

### Preface
It happens five to ten times every year that questions like these reach us: 'There is a grub present in this product, of which insect is it the larva?' Or: 'I want to know which kind of insect this is and is it harmful?'

### Method
*Larvae*: they are hard to tell apart, so we put them in a jar with something to eat and wait for them to change to the imago.
*Insects*: we look at them closely with the aid of a stereomicroscope and compare them with pictures and descriptions in literature.

### Literature
(1) Walther Faber: *Wichtige Vorratsschädlinge und ihre Bekämpfung*.
(2) Hermann Bollow: *Vorrats- und Gesundheitsschädlinge*.
(3) *Elseviers gids van nuttige en schadelijke insecten*.
(4) Jiry Zahradnik: *A field guide in colour to insects*.

### A useful table

| Family | Species | Found in |
|---|---|---|
| mites | flourmite | cereal |
| psocids | dustlouse | dried mushrooms |
| cockroaches | American cockroach | bread |
| moths | Mediterranean flour moth | flour |
| beetles | mealworm beetle | rolled oats |
| flies | flesh-fly | flesh |
| ants | Pharoah ant | dried rasins |
| wasps | common wasp | marmalade |

### In practice
What have we learnt in the last couple of years?

50 per cent of the questions concern beetles, 40 per cent moths and 10 per cent the rest. The product that is most often contaminated with insects is flour. The number of questions about insects in products is about equal to the number of questions about insects found about the house and thought to be dangerous for foodstuffs or furniture. Some examples: ground beetles in the cellar of a Chinese restaurant; great brown weevils in a living room; longhorn beetles coming out of firewood.

## VI.36   A DISTRIBUTION IN CLASSES FOR AN OBJECTIVE QUALITY ASSESSMENT OF SHRIMPS

**Th. Croegaert**
SGS-FAC, AGRILAB®, Polderdijkweg (Hansadok 407), B-2030 Antwerp, Belgium
**R. Vanthuyne**
Laboratory for Chemical and Physical Analysis of Fish (for the purpose of Veterinary Expertise), Ostend, Belgium
**J. Van Hoof**
Laboratory for Hygiene and Technology of Food from Animal Origin, Faculty of Veterinary Medicine, State University of Ghent, Belgium

The inspection of fish and shellfish relates to freshness, undamaged condition, sanitary soundness and temperature, and must lead to the decision whether the product is suitable or not for human consumption.

Several authors examined shrimps sold on the Belgian market. For this purpose, they used microbiological, physical, chemical and organoleptic analytical techniques. From these studies, it appeared that in general the analysed shrimps were of rather poor consumption quality. It was also obvious that microbiological analysis was not relevant for the assessment of the consumption quality, since several antimicrobial treatments are used in shrimp processing (boiling, irradiation, addition of acids and disinfectants). Chemical and physical analyses, which are better related to organoleptic characteristics, should provide decisive information.

Evidence was given of the necessity of applicable standards, which allow an objective quality judgement for shrimps. These standards should take account of the raw or cooked condition of shrimps. Finally, sampling plans were to be elaborated, to permit an objective judgement of a whole consignment by means of a representative sample.

In this contribution, an examination plan is presented which allows a total quality evaluation. Both the freshness and the hygienic condition of the shrimps are taken into account.

Freshness of the shrimp is interpreted in terms of 'tropical' quality (indol-concentration), chemical quality (total volatile basic nitrogen- and ammonia-concentration) and physical quality (pH value).

The hygienic condition of the shrimp, as an index for sanitary soundness, is evaluated in regard to total plate counts, to sewage bacteria (Enterobacteriaceae, coliforms, faecal streptococci) and to pathogens (*Staphylococcus aureus, Salmonella, Shigella, Vibrio parahaemolyticus*, sulphite reducing clostridia).

For the tropical, the physical/chemical and the hygienic qualities different standards are proposed for raw and cooked shrimps. For each criterion three scores can be attributed in terms of 'good', 'sufficient' and 'insufficient'. In this way, shrimp samples can be put into different quality classes. A sampling plan using five samples is advised in order to permit a statistically relevant judgement.

For the practical evaluation of this proposal, analytical results of 143 samples were judged according to the given standards. From this comparison, evidence was given that 40 per cent of raw tropical shrimps, 60 per cent of cooked tropical shrimps and 82 per cent of cooked brown shrimps were of 'insufficient quality' and therefore unsuitable for human consumption.

# Index